Microwave Heating - Electromagnetic Fields Causing Thermal and Non-Thermal Effects

Edited by Gennadiy I. Churyumov

Published in London, United Kingdom

IntechOpen

Supporting open minds since 2005

Microwave Heating - Electromagnetic Fields Causing Thermal and Non-Thermal Effects
http://dx.doi.org/10.5772/intechopen.87921
Edited by Gennadiy I. Churyumov

Contributors
Gaoming Lu, Maria Zaharescu, Szilagyi Imre Miklos, Luminita Predoana, Irina Stanciu, György Pokol, Dániel Attila Karajz, Vincent Otieno Odhiambo, Ravindra Kamble, Sheetal Marganakop, Pramod Kattimani, Sudha Belgur Satyanarayana, Tetyana Frolova, Gennadiy I. Churyumov, Vyacheslav Buts, Eugene Odarenko, Vladimir Gerasimov, Simona Collina, Pasquale Linciano, Valeria Cavalloro, Emanuela Martino, Bikas C. Das, Sandhya K. M, Litty Thomas Manamel, Abdurahman Nour, Oluwaseun Alara, N.H. Azhari, Manal Omer, Noormazlinah Ahmad, Akira Naito, Yugo Tasei, Batsaikhan Mijiddorj, Izuru Kawamura, Kazuyoshi Ueda, Jianjun Zhou

First published in London, United Kingdom, 2021 by IntechOpen
IntechOpen is the global imprint of INTECHOPEN LIMITED, registered in England and Wales, registration number: 11086078, 5 Princes Gate Court, London, SW7 2QJ, United Kingdom
Printed in Croatia

British Library Cataloguing-in-Publication Data
A catalogue record for this book is available from the British Library

Additional hard and PDF copies can be obtained from orders@intechopen.com

Microwave Heating - Electromagnetic Fields Causing Thermal and Non-Thermal Effects
Edited by Gennadiy I. Churyumov
p. cm.
Print ISBN 978-1-83968-226-1
Online ISBN 978-1-83968-227-8
eBook (PDF) ISBN 978-1-83968-228-5

For EU product safety concerns:
IN TECH d.o.o., Prolaz Marije Krucifikse Kozulić 3, 51000 Rijeka, Croatia, info@intechopen.com or visit our website at intechopen.com.

Meet the editor

Gennadiy I. Churyumov was born on February 12, 1952, in the former USSR. He received a Dipl.-Ing. in Electronic Engineering and a Ph.D. in Radiophysics from the Kharkiv Institute of Radio Electronics, Kharkiv, Ukraine, in 1974 and 1981, respectively. In 1997, he received a DSc from the Institute of Radio Physics and Electronics of the National Academy of Sciences of Ukraine. Since 2002, he has been a professor at the Kharkiv National University of Radio Electronics, where he also is a scientific adviser in the Microwave & Optoelectronics Lab. From 2002 to 2018, he served as editor in chief of the International Journal on Applied Radio Electronics. Since 2016, he has been a professor in the Electronics Engineering Department, Harbin Institute of Technology, China. His personal research interests include the theory and simulation of electromagnetic problems, microwave vacuum electron tubes, and practical aspects of electromagnetic energy application. He has published more than 320 research articles including books, book chapters, journal papers, international conference proceedings, and patents. He is a senior member of the Institute of Electrical and Electronics Engineers (IEEE).

Contents

Preface

Microwave heating is the process whereby the electromagnetic field of the microwave range (300–3000 MHz) interacts with diverse dielectric materials and mediums that take up electromagnetic energy. In practice, the most popular frequencies for microwave heating are 915 and 2450 MHz. The process of absorption and transformation of electromagnetic energy inside a volume of various objects results in self-heating (or in a manifestation of the so-called thermal effect). The energy efficiency of the given process depends on both the dielectric properties of the materials (specific conductivity, loss factor, water content, etc.) and the parameters of the electromagnetic field (as a rule, they are the amplitude, the frequency, and the operation modes—continuous or pulse). It should be noted that in the case of microwave heating, the impact of electromagnetic energy on objects takes place at the molecular level. This is its main feature compared to conventional heating techniques. This property of microwave heating has advantages associated with uniformity, efficiency, higher speed, and ecological compatibility.

More than 80 years of experience in the practical application of electromagnetic energy in various fields of human activity (including industry, agriculture, science, medicine, etc.) suggests that microwave heating is an effective application of the energy of the electromagnetic field. Essentially, new microwave technology was created to solve numerous problems in diverse applied fields. The modern technology of microwave heating is constantly evolving. Primarily, this is associated with further enhancing the efficiency of microwave heating and applying the thermal effect to the solution of problems in microwave chemistry (so-called green chemistry), producing new materials possessing new properties, plasma physics, and more. It is important to pay attention to the questions related to the nonthermal microwave effect, taking into account that the mechanism of the given effect has not yet been sufficiently studied and understood. Based on the available information, the nonthermal effect of microwave irradiation (usually, in the range of 40–60 GHz) can be defined as information impact. This process occurs on a cellular level with a subsequent impact on the macro parameters of all considered systems as a whole. As a rule, such an impact regime of microwave irradiation is used in medicine for treating some human diseases as well as in biology and virology for changing the conditions of existing microorganisms (viruses, bacteria, etc.).

This book presents the latest investigations related to the application of microwave energy in various fields. It is divided into two sections on thermal and non-thermal effects of microwave radiation on materials and mediums. The book includes eight chapters written by various authors presenting academic and industrial interests. Chapters 1 and 2 focus on the advantages of using microwave energy for the extraction (recovery) of bioactive compounds from plant materials. Chapter 3 reviews recent developments in the microwave-assisted synthesis of functional dyes for subsequent use in hi-tech applications, such as optoelectronics, photochromic materials, liquid crystal displays, newer emissive displays, biomedicine, electronic materials (organic semiconductors), and imaging technologies. Chapter 4 discusses the effective use of microwave heating for high-quality nanomaterial synthesis in solutions where doping is necessary to tune the electronic and optoelectronic properties of nanomaterials used in various applications. Chapter 5 examines the influence of microwaves on the chemical reactions that take place during sol-gel synthesis as well as on the properties of the resulting samples. Chapters 6 and 7 study the influence of microwaves on gaseous mediums and hard materials (rocks). Chapter 6 focuses on the experimental and theoretical results of exciting a low-temperature plasma with the help of a gaseous medium by both the regular field (in a regime with dynamic chaos) and random electromagnetic fields. The chapter also reviews and analyzes the stochastic heating mechanisms of plasma-derived for greater efficiency. Chapter 7 generalizes a practical experience of the application of microwaves for rock-breaking operations. It presents a mechanism of the rock-breaking operation concerning the heating characteristics of the rocks and their mechanical properties. Chapter 8 focuses on the influence of microwaves on the chemical reactions occurring in situ at a molecular level using microwave irradiation nuclear magnetic resonance spectroscopy and molecular dynamics simulation. It shows that the chemical reaction rate can be sped up due to the ordered state of the polar molecules that were induced by a nonthermal effect.

I would like to thank all the authors for their excellent contributions. I would also like to take this opportunity to thank Author Service Manager Ms. Sara Debeuc at IntechOpen for her efforts and help.

Gennadiy I. Churyumov
Professor,
Antennas and Microwave Engineering Center,
Microwave Engineering Department,
Harbin Institute of Technology,
China

Professor,
Department of Physical Foundation of Electronic Engineering,
Faculty of Electronic Engineering,
Kharkiv National University of Radio Electronics,
Ukraine

Section 1

Thermal Effect of Electromagnetic Field

Chapter 1

Microwave-Assisted Extraction of Bioactive Compounds (Review)

Abdurahman Hamid Nour, Alara Ruth Oluwaseun, Azhari Hamid Nour, Manal Suliman Omer and Noormazlinah Ahmed

Abstract

In recent times, bioactive compounds from plant samples are extracted using a microwave extractor. This is because traditional methods of extraction are need of higher volume of solvents, degrade thermal-sensitive bioactive compounds, and consume much time of extraction. Hence, this chapter unveils the importance of the microwave-assisted extraction (MAE) technique in the recovery of bioactive compounds from plants. The involving extraction steps need to recover higher yields, faster, consumption of lesser extracting solvents, and ensure stable heat-sensitive bioactive compounds. The factors affecting MAE in the recovery of bioactive compounds from plant materials are as well discussed. Additionally, some of the previously reported bioactive compounds from plant samples using MAE are highlighted.

Keywords: extraction, microwave heating, microwave-assisted extraction, bioactive compounds, solvents, plants

1. Introduction

Extraction involves separating dissolvable substances from non-dissolvable residues using solvent(s); it can be in form of liquid or solid [1]. There are two categories of extraction which are traditional and modern; the former includes Soxhlet, soaking, maceration, ultra-sonication, turbo-fast blending, and solvent permeation; the latter includes ultrasonic-assisted, subcritical, supercritical CO_2, enzyme-assisted, pressure-assisted, and microwave-assisted methods [2–6]. The traditional methods are mainly associated with an extended time of extraction, destruction of heat-sensitive bioactive compounds, and enormous consumption of solvents [3, 7]. It is then important to explore modern methods of extraction to overcome the setbacks associated with the traditional methods. Out of all the modern methods of extraction, microwave-assisted extraction (MAE) has received the greatest attention due to its reduced consumption of solvent, shorter operation time, reproducibility, improved recovery yield, good selectivity, and reduced sample manipulation [8, 9]. Gedye et al. and Giguere et al. were groups that first described the usage of microwave energy in 1986, it was employed in organic synthesis; microwave energy was also employed in the extraction of biological samples for analyzing organic compounds [10–12].

MAE method is being used in different kinds of samples which include geological, environmental, and biological matrices. In recent times, MAE is generally used

Solvent	Dielectric loss	Dielectric constant	Loss tangent
Chloroform	0.437	4.8	0.091
Dimethyl sulfoxide	37.125	45.0	0.825
Dimethylformamide	6.079	37.7	0.161
Ethanol	22.866	24.3	0.941
Ethylene glycol	49.950	37.0	1.350
Hexane	0.038	1.9	0.020
Toluene	0.096	2.4	0.040
Water	12.3	80.4	9.889

Table 1.
Solvents with their corresponding dielectric losses, dielectric constants, and loss tangents.

in obtaining bioactive compounds from plant samples, this has greatly improved the total interest in development and research areas. This method allows for faster recovery of solutes from plant samples with appreciable extraction efficiency as compared to traditional techniques. MAE is one of the modern methods, and employed shortened time of extraction, minimal solvent consumptions, and secure thermolabile compounds. It is a green technology that is effective for extracting bioactive compounds from plant samples [13]. Based on the importance of MAE, this method has provided two sub-classes which are microwave solvent-free extraction (MSFE) and microwave-assisted solvent extraction (MASE).

Microwave irradiation employs a specific frequency of electromagnetic field in a way closely to photochemical-activated reaction; the frequency falls between 300 MHz and 300 GHz [14]. Nevertheless, few frequencies are allowed for medical, scientific and industrial usages; this falls within 0.915 and 2.45 GHz worldwide. Dielectric heating from MAE is appropriate for heat-sensitive bioactive compounds [15]. It had been provided that the used water for extracting phenolic compounds is not effective compared to traditional techniques due to reduced dissipation factor and higher dielectric constant associated with water relative to other solvents; hence, using solvents that possess higher dissipation and dielectric factors is advisable in MAE. Furthermore, extractability is proportional to the solvent used in extracting bioactive compounds from plants and kind of plant sample [16]. **Table 1** presents the dielectric losses, dielectric constants, and loss tangents for different solvents used in MAE. Rapid heating is generated in MAE when ionic species or polar molecules are used, this heating generates collisions with molecules from surrounding which do not require higher pressure. In most cases, the extraction time and microwave power fall within 30 s to 10 min and 25 to 750 W, respectively [17]. Several studies had reported the use of MAE for recovering phenolics from plant samples including bitter leaf, purple fleabane, roselle, tea leaf, vanilla, radix, flax seeds, scent leaf, siam weed, and among others [6, 8, 9, 18–22].

Thus, the chapter presents the working principle, factors influencing this method, and previously reported bioactive compounds extracted through MAE.

2. Operating principle and working mechanism of MAE

2.1 Operating principle of MAE

The fundamental of MAE technique is different compared to traditional techniques, this is because MAE happens based on electromagnetic waves that causes the cell structure to change. Microwave-assisted extraction works with a principle by which polarizable materials and dipoles of polar solvent interact with microwave radiation whereby the forces between magnetic and electric components change direction rapidly. The molecules of polar solvent get heated when they orient in the changing field direction. In the case of non-polar solvents that do not have polarizable groups, the heating is poor. This thermal effect at the molecular level is rapid but limited to the depth near the surface and a small portion of the samples. The remaining part of the samples is heated up by conduction. Therefore, this is the major drawback of the MAE because large samples or agglomerates of small samples cannot be heated uniformly. There is a possibility of using high power sources in order to enhance the depth of penetration but microwave radiation involves an exponential decay once inside a microwave-absorbing solid [23].

2.2 Working mechanism of MAE

The mechanism at which microwave-assisted extraction works is different from other types of extraction methods because the extraction occurs as a result of changes in the cell structure caused by electromagnetic waves [3]. As provided in **Figure 1**, this process of extraction involves a synergistic combination of mass and heat transfers working in the same direction whereas the mass transfer in conventional methods occurs from inside to outside of the substrates and heat transfer occurs from the outside to inside of the substrate [13]. The series of phenomenological steps that occur during the microwave-assisted extraction (MAE) are as follows:

a. The irradiation heat from a microwave is transferred to the solid through the microwave-transparent solvent without absorption;

b. The intense heating of the (a) above results in residual microwave-absorbing in the solid being heated up;

c. The heated moisture evaporates and creates a high vapor pressure;

d. The high vapor pressure breaks the cell of the substrate; and

e. Cell wall breakage enhances the releases of the extract from the samples [13].

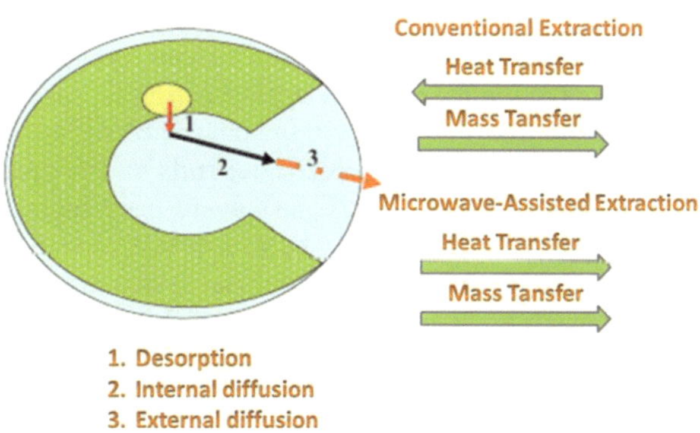

Figure 1.
Heat and mass transfer mechanisms in conventional and microwave extraction [13].

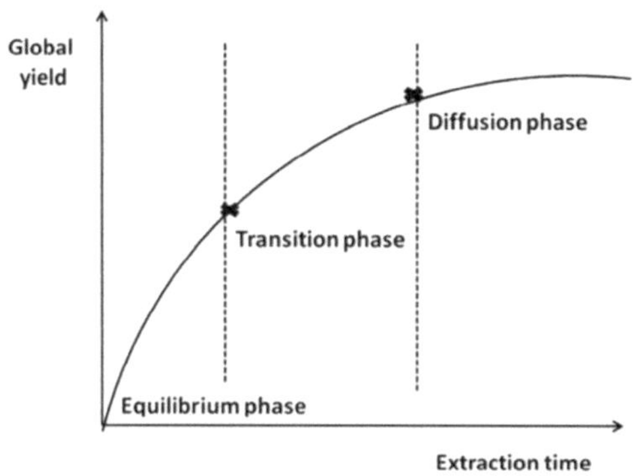

Figure 2.
Pictorial diagram of yield against time in the extraction [14].

Additionally, the extracting solvent is absorbed into the plant sample through diffusion, causing the dissolution of solutes into the solvent until saturation. This solution diffuses to the plant surface through effective diffusion and then transfer to the bulk solution (**Figure 2**). Several forces that include physicochemical relations and interactions can be seen during the process (chemical interactions, driving forces, interstitial diffusion, and dispersion forces), and the strength and persistence of properties can be related to the characteristics of the extraction solvent (polarity, solubility in water, purity, solubilization, and among others) [4].

3. Essential factors influencing MAE and mechanism of action

Several studies had been done on optimizing MAE factors to achieve optimal yields from the considered plant samples. The operative parameters influencing MAE include solvent-to-feed ratio, solvent composition, characteristic of the plant sample and its water content, microwave power, irradiation time, stirring effect, microwave energy density, and extraction temperature. These operative parameters determine the efficiency of MAE. Hence, understanding the influences and interactions of these parameters on the extraction process is paramount.

3.1 Solvent-to-feed ratio

The selection of solvent is the most significant factor that affects microwave-assisted extraction. Adequate solvent selection will produce an efficient extraction process. The solubility of the compound of interest, mass transfer kinetics of the process, and solvent penetration that occurs from the interaction between the dielectric effect and sample matrix are inevitable parameters [24, 25]. Chan et al. reported that the selection of extraction solvent depends on the capacity of that solvent to absorb microwave energy [26]. If the solvent has a high dielectric constant and dielectric loss, the solvent capacity to absorb microwave energy will be high [25]. Tatke and Jaiswal reported that solvents such as methanol, ethanol, and water are excellent microwave-absorbing solvents which possess sufficient polarity to be heated up through microwave power [27]. Studies had shown that the addition of a small quantity of water to polar solvent resulted in higher diffusion of water into the cells of the matrix, leading to effective heating and thus facilitating the transport of compounds into the solvent at higher mass transfer rates [24, 26, 28].

Veggi et al. had reported that the extraction solution must not exceed 30–34% (w/v) [29]. In the past studies, the solvent-to-feed ratio between 10:1 (mL/g) and 20:1 (mL/g) had been reported to give optimal yields [29, 30]. The volume of extracting solvent is another important factor, a large volume of solvent requires more energy and time to condense extraction solution in the purification process. MAE may give lower recoveries because of non-uniform distribution and exposure to microwave [29].

3.2 Irradiation time

The irradiation time is another important factor that affects microwave-assisted extraction. One of the importance of MAE over conventional methods is that the extraction time is very short. The usual time ranges from a few minutes to half an hour depending on the plant matrix so as to avoid possible oxidation and thermal degradation [13, 25, 27]. The irradiation time is affected by the dielectric property of solvent used. Solvents such as ethanol, water, and methanol may heat up rapidly on longer exposure which can result in degradation of thermolabile compounds in the extracts [4, 26]. Increased time of irradiation can improve the recovery yield; nevertheless, the increased yield can decline at prolonged irradiation time [21].

Sometimes, if the extraction will take a longer time, the plant materials are extracted through multiple stages by utilizing consecutive extraction cycle. Here, a new solvent is introduced to the residues, the procedure is then repeated to ensure exhaustion of the plant sample. The use of this process helps higher recovery yield with no excessive heating [26, 31]. The nature of plant sample and solute determines the number of extraction cycles. A study presented that 3 cycles of 7 min were adequate in extracting triterpene saponins from yellow horn through MAE [32]. The optimization MAE to obtain triterpenoids saponins from *Ganoderma atrum* yielded 5 min for each cycle [33].

3.3 Effect of stirring

Mass transfer processes in the solvent phase are usually enhanced by stirring. The equilibrium between the vapor and aqueous phases is achieved more rapidly. The use of a stirrer in MAE accelerates the extraction process by increasing the dissolution and desorption of bioactive compounds in the sample matrix [13, 27]. Thorough stirring can reduce the drawbacks possess when using a low solvent-to-solid ratio and minimized the mass transfer barrier [13].

3.4 Microwave power and temperature

Microwave power and temperature are important factors that affect the extraction yield when using MAE. The higher microwave power can lead to an increase in the temperature of the system resulting in the increase of the extraction yield until it becomes insignificant or declines [13, 25, 34]. An increase in temperature can result in solvent power increase because of a drop in surface tension and viscosity, enhancing the solvent to solubilize solutes, improving matrix wetting and penetration [13]. However, Spigno and De Faveri reported that the efficiency of MAE increases with the increase in temperature until an optimum temperature is reached [25]. Microwave power is also related to the quantity of sample and the extraction time required. However, the power provides localized heating in the plant matrix acts as a driving force for MAE to destroy the plant matrix so that the solute can diffuse and dissolve in the solvent. Therefore, increasing the microwave power will generally improve the extraction yield and result in a shorter extraction time [13, 29, 35]. On the other hand, if microwave power is too high, it can result in poor extraction yield leading to the degradation of thermally sensitive compounds in the plant matrix [29]. It is then important to select the appropriate microwave power to reduce the extraction time required to reach the set temperature and avoid a "bumping" phenomenon [13].

3.5 Characteristic of plant sample and its water content

The characteristic of the plant sample and its water content can influence MAE. The extraction efficiency improves as the contact surface area of the plant sample increases. Moreover, finer samples give room for deeper penetration of microwave irradiation [36]. Nevertheless, too much finest of the plant sample may generate some technical difficulties; hence, filtration or centrifugation is employed in the preparation of the plant samples [27, 37]. During the sample preparation, the grinded sample is homogenized to improve contact between the solvent and the plant matrix. The plant particle sizes mostly fall within 2 and 100 mm [31]. Sometimes, the plant matrix is soaked before extraction to improve the yield; this is known as pre-leaching [37].

Mostly, the recovery of bioactive compounds from the plant matrix tends to increase through its moisture that acts as a solvent. This moisture is heated up, evaporated, causes pressure within the cell, and dispenses the solutes through rupturing of the cell wall; thus, increase the yield of bioactive compounds [38]. An increase in the polarity of solvent causes the addition of water to have a positive influence on microwave-absorbing capability; thus, encourages the heating procedure [26]. Extra water generates hydrolyzation and reduces the oxidation of bioactive compounds.

3.6 Microwave energy density

There are three heating operational modes employed in the performance evaluation of microwave-assisted extraction [28]. These include the constant-power heating mode, intermittent heating mode, and the constant temperature heating mode. Terigar et al. reported that the constant power heating mode presents the standard practice in the extraction of thermally sensitive active constituents of the plant matrix [35]. It is worthy to note that the microwave power alone does not provide an adequate explanation as to how energy is being absorbed in the extraction of the biological medium. Li et al. therefore studied the interrelationship between the microwave energy density and the extraction yield, it was concluded that for a unit of extracting solvent, microwave energy density is the most important factor affecting the extraction efficiency in a microwave-assisted extraction [39].

Gao et al. reported an accelerated effect on the ionic conduction and dipole rotation which in turn leads to an increase in the extraction yield [40]. This is due to the release of more microwave energy to the biological medium as the microwave power increases. Polar solvents rates of absorption improve with increasing power and ultimately resulting in higher heating and extraction rate [41]. Li et al. in [39] described the energy density of microwave heating as the power per unit quantity of sample under extraction as shown in Eq. (1).

$$Energy\ density\,(W/mL) = \frac{Microwave\ power\,(W)}{Volume\ of\ extracting\ solvent\,(mL)} \quad (1)$$

Number	Plant sample	Results obtained	Bioactive compounds	Remarks	Reference
1.	*Artemisia annua* L.	MAE: Microwave power = 650 W; Solvent/feed ratio = 15; Temperature = ambient; Extraction time = 12 min SFE: Pressure = 30 MPa; solvent = CO_2; Solvent/feed ratio = 6; Temperature = 35 °C; Extraction time = 2.5 h Soxhlet: Solvent oil; S/F = 11.67; T = 35 C; t = 6 h	Artemisinin (92.1% db) Artemisinin (33.2% db) Artemisinin (60.4% db a)	High yields and selectivity compared to other extraction methods.	[43]
2.	Sweet grass leaves	MAE: Microwave power = 200 W; solvent used = acetone; Solvent/feed ratio = 10; Temperature = 80 °C; Extraction time = 15 min; one-step extraction SFE: Two-step: 1. Pressure = 35 MPa; Temperature = 40 °C 2. Pressure = 25Mpa; Temperature = 40 °C; Solvent = 20% of ethanol; Extraction time = 2 h; Flowrate = 0.5 L/min Soxhlet: Solvent/feed ratio = 50; Solvent = acetone; Extraction time = 6 h	5,8-Dihydroxycoumarin (0.42% db) 5-Hydroxy-8-*O*-β-D-glucopyranosyl-benzopyranone (0.11% db) 5,8-Dihydroxycoumarin (0.49% db) 5-Hydroxy-8-*O*-β-D-glucopyranosyl-benzopyranone (0.06% db) 5,8-Dihydroxycoumarin (0.46% db) 5-Hydroxy-8-*O*-β-D-glucopyranosyl-benzopyranone (0.08% db)	High yields and selectivity compared to other extraction methods.	[44]
3	Licorice roots	MAE: Microwave power = 700 W; Solvent = ethanol; Solvent/feed ratio = 10; Temperature = 85–90 °C; Extraction time = 4 min Ultrasonic: Solvent = ethanol; Solvent/feed ratio = 10; Extraction time = 20.5 h Soxhlet:	Glycyrrhizic acid–GA (2.26%) Glycyrrhizic acid–GA (2.26%) Glycyrrhizic acid–GA (2.5%)	It recovered a higher yield in reduced time.	[45]

Number	Plant sample	Results obtained	Bioactive compounds	Remarks	Reference
		Solvent = ethanol; Solvent/feed ratio = 10; Extraction time = 10 h			
4.	Green tea leaves	MAE: Microwave power = 700 W; Solvent = ethanol/water (1:1 v/v); Solvent/feed ratio = 20; Temperature = 20 °C; Extraction time = 4 min UAE: Solvent = ethanol/water (1:1 v/v); Solvent/feed = 20; Temperature = 20–40 °C; Extraction time = 90 min Heat reflux extraction: Solvent = ethanol/water (1:1 v/v); Solvent/feed = 20; Temperature = 85 °C; Extraction time = 45 min	Tea polyphenols (30%), Tea caffeine (4%) Tea polyphenols (28%), Tea caffeine (3.6%) Tea polyphenols (28%), Tea caffeine (3.6%)	High yields and selectivity compared to other extraction methods.	[18]
5.	Grape fruit	MAE: Microwave power = 0.9 kW; Solvent = water; Solvent/feed ratio = 30; T = 20 °C; Extraction time = 6 min UAE: Solvent = water; Solvent/feed ratio = 30; T = 70 °C; Extraction time = 25 min UAE + MAE: Microwave power = 0.45 kW; Solvent/feed = 30; Extraction time = 30 min for UAE and 10 min for MAE Heat batch: Solvent = water; Solvent/feed = 30; T = 90 °C; Extraction time = 90 min	Pectin (27.81%) Pectin (17.92%) Pectin (31.88%) Pectin (19.16%)	High yields compared to other extraction methods.	[46]
6.	*Ganoderma atrum*	MAE: Solvent = ethanol/water (9.5:0.5 v/v); Solvent/feed ratio = 25; Temperature = 90 °C; Extraction time = 5 min	Global yield (5.11% db)	High yields compared to other extraction methods.	[47]

Number	Plant sample	Results obtained	Bioactive compounds	Remarks	Reference
		UAE: Solvent = ethanol/water (9.5:0.5 v/v); Solvent/feed ratio = 25; Extraction time = 30 min; Frequency = 33 kHz	Global yield (1.72% db)		
		SFE: Pressure = 25 MPa; Temperature = 55 °C; Solvent = CO_2 + ethanol; Extraction time = 3 h	Global yield (1.52% db)		
		Shaking: Solvent = ethanol/water (9.5:0.5 v/v); Extraction time = 3 h	Global yield (2.58% db)		
		HRE: Solvent = ethanol/water (9.5:0.5 v/v); Solvent/feed ratio = 25; Temperature = 95 °C; Extraction time = 1 h	Global yield (2.22% db)		
7.	Yellow hcrn	MAE: Microwave power = 900 W; Solvent = ethanol/water (40:60 v/v); Solvent/feed = 30; Temperature = 50 °C; Extraction time = 7 min × 3 cycles	Global yield (11.62%)	High yields compared to other extraction methods.	[48]
		UAE: Microwave power = 250 W; Solvent = ethanol/water (40:60 v/v); Solvent/feed ratio = 30; Temperature = 50 °C; Extraction time = 60 min × 3 cycles	Global yield (6.78% db)		
		HRE: Microwave power = 800 W; Solvent = ethanol/water (40: 60 v/v); Solvent/feed ratio = 30; Temperature = 50 °C; Extraction time = 90 min × 3 cycles	Global yield (10.82% db)		
8.	Turmeric plant	MAE: Microwave power = 60 W; Solvent = acetone; Solvent/feed = 3; Temperature = 50 °C; Extraction time = 5 min	Curcumin (90.47% db)	High yields compared to other extraction methods.	[49]

Number	Plant sample	Results obtained	Bioactive compounds	Remarks	Reference
		UAE: Microwave power = 150 W; Solvent = acetone; Solvent/feed = 3; Temperature = 21 °C; Extraction time = 5 min	Curcumin (71.42% db)		
		Soxhlet: Solvent = acetone; Solvent = 5; Extraction time = 8 h	Curcumin (2.10% db)		
		SFE: Pressure = 30 MPa; Solvent = CO_2 + ethanol (10%); Temperature = 50 °C; Extraction time = 240 min; flowrate = 5 mL/min	Curcumin (69.36% db)		
9.	*Silybum marianum* (L.) (milk thistle)	MAE: Microwave power = 600 W; Solvent = ethanol/water (80:20 v/v); Solvent/feed = 25; Extraction time = 2 min × 6 cycles	Silybinin (1.37 db)	High yields compared to other extraction methods.	[50]
		Soxhlet: Solvent = ethanol/water (80:20 v/v); Solvent/feed = 100; Extraction time = 12 h	Silybinin (1.09 db)		
		Stirring: Solvent = ethanol/water (80:20 v/v); Solvent/feed ratio = 100; Extraction time = 24 h	Silybinin (0.48% db)		
		Maceration: Solvent = ethanol/water (80:20 v/v); Solvent/feed ratio = 100; Extraction time = 24 h	Silybinin (0.36 db)		
10.	*Coriandrum sativum*	MAE: Microwave power = 200 W; Solvent = ethanol/water (50:50 v/v); Solvent/feed = 20; Temperature = 50 °C, Extraction time = 18 min	Phenolics content (0.082% db)	The recovery of phenolic compounds was higher in MAE compare to other techniques.	[51]
		UAE: Solvent = ethanol/water (50:50 v/v), Solvent/feed ratio = 10; Extraction time = 30 min	Phenolics content (0.041% db)		

Number	Plant sample	Results obtained	Bioactive compounds	Remarks	Reference
11.	*Cinnamomum zeylanicum*	MAE: Microwave power = 200 W; Solvent = ethanol/water (50:50 v/v); Solvent/feed ratio = 20; Temperature = 50 °C, Extraction time = 18 min UAE: Solvent = ethanol/water (50:50 v/v), Solvent/feed ratio = 10; Extraction time = 30 min	Phenolics content (1.679% db) Phenolics content (0.506% db)	The recovery of phenolic compounds was higher in MAE compare to other techniques.	[51]
12.	*Cuminum cyminum*	MAE: Microwave power = 200 W; Temperature = 50 °C, Solvent = ethanol/water (50:50 v/v); Solvent/feed ratio = 20; Extraction time = 18 min UAE: Solvent = ethanol/water (50:50 v/v), Solvent/feed ratio = 10; Extraction time = 30 min	Phenolics content (1.159% db) Phenolics content (0.290% db)	The recovery of phenolic compounds was higher in MAE compare to other techniques.	[51]
13.	*Crocus sativus*	MAE: Microwave power = 200 W; Temperature = 50 °C, Solvent = ethanol/water (50:50 v/v); Solvent/feed ratio = 20; Extraction time = 18 min UAE: Solvent = ethanol:water (50:50 v/v), Solvent/feed ratio = 10; Extraction time = 30 min	Phenolics content (2.939% db) Phenolics content (0.500% db)	The recovery of phenolic compounds was higher in MAE compare to other techniques.	[51]
14.	Sea buckthorn	MHG: Microwave power = 400 W; Extraction time = 15 min; Humidity = 57% Agitated: Solvent = methanol/water (80:20 v/v);	Isorhamnetin 3-*O*-rutinoside (0.123% db) Isorhamnetin 3-*O*-glucoside (0.097% db) Quercetin 3-*O*-Glucoside (0.025% db) Isorhamnetin (0.00084% db) Isorhamnetin 3-*O*-rutinoside (0.187% db)	Recovery of higher yields of bioactive compound compared to other extraction techniques.	[52]

Number	Plant sample	Results obtained	Bioactive compounds	Remarks	Reference
		Solvent/feed ratio = 10; Extraction time = 8 min	Isorhamnetin 3- *O* -glucoside (0.162% db) Quercetin 3- *O* -Glucoside (0.016% db) Isorhamnetin (0.00064% db)		
15.	Cranberry press cake	MAE: Solvent = ethanol; Solvent/feed ratio = 5.7; Temperature = 125 °C; Extraction time = 10 min Stirring: Solvent = ethanol; Solvent/feed ratio = 5; Extraction time = 2 h	Quercetin (0.1537% db) Quercetin (0.1272% db)	Recovery of higher yields of bioactive compound in lesser time compared to other extraction techniques.	[53]
16.	*Morinda citriflora* (roots)	MAE: Microwave power = 720 W; Solvent = ethanol/ water (80:20 v/v); Solvent/feed ratio = 100; Temperature = 60 °C; Extraction time = 15 min UAE: Solvent = ethanol; Solvent/feed ratio = 100; Temperature = 60 °C; Extraction time = 60 min Maceration: Solvent = ethanol; Solvent/feed ratio = 100; Extraction time = 3 days Soxhlet: Solvent = ethanol; Solvent/feed ratio = 100; Temperature = 100 °C; Extraction time = 4 h	Global yield (95.91% db) Global yield (62.23% db) Global yield (63.33% db) Global yield (97.74% db)	High yields compared to other extraction methods.	[54]
17.	Soybean germ	MAE: Solvent/feed ratio = 17.5; Temperature = 120 ° C; Extraction time = 0.5 h MAE + UAE: Microwave power = 60 W for UAE and 100 W for MAE; Solvent/feed ratio = 5;	Global yield (16.5% wb) Global yield (14.1% wb)	High yields compared to other extraction methods.	[55]

Number	Plant sample	Results obtained	Bioactive compounds	Remarks	Reference
		Temperature = 45 °C; Extraction time = 1 h Soxhlet: Solvent = hexane; Solvent/feed ratio = 6.67; Extraction time = 4 h	Global yield (8.65% wb)		
18.	*Lavandula angustifolia* Mill., Lamiaceae (lavender flowers)	MASD: Microwave power = 500 W; Solvent = water; Solvent/feed ratio = 4; Extraction time = 10 min SD: s = water; S/F = 4; t = 90 min	Monoterpenes (3.45% db) Oxygenated monoterpenes (78.29% db) Sesquiterpenes (2.77% db) Global yield (8.86% db) Monoterpenes (4.92% db) Oxygenated monoterpenes (75.14% db) Sesquiterpenes (2.87% db) Global yield (2.59% db)	Recovery of higher yields of a bioactive compound in lesser time compared to other extraction techniques.	[56]
19.	Caraway (*Carum carvi* L.)	MDG: Microwave power = 100 W; Extraction time = 45 min Hydrodistillation: Solvent/feed ratio = 5; Extraction time = 300 min	Global yield (2.59% db) Carvone (67.59% db) Limonene (30.10% db) Global yield (2.54% db) Carvone (66.89% db) Limonene (30.30% db)	High yields compared to other extraction methods.	[57]
20.	Tomato	MAE: Microwave power = 100 W; Solvent = methanol; Solvent/feed ratio = 50; Extraction time = 45 min Shaker: Solvent = ethanol/water (60:40 v/v); Solvent/feed ratio = 50; Temperature = 45 °C; Revolution = 400 rpm; Extraction time = 15 h	Total phenolic contents (0.646% db) Total phenolic contents (0.603% db)	The recovery of phenolic compounds was higher in MAE compare to other technique.	[58]
21.	*Foeniculum vulgare* Miller (seeds)	MWHD: Microwave power = 300 W; Solvent = water; Solvent/feed ratio = 2; Temperature = 100 °C; Extraction time = 200 s	Global yield (1.14% db)	High yields compared to other extraction methods.	[59]

Number	Plant sample	Results obtained	Bioactive compounds	Remarks	Reference
		HD: Microwave power = 300 W; Solvent = water; Solvent/feed ratio = 8; Extraction time = 319 s; Temperature = 100 °C; Revolution = 50 rpm	Global yield (0.265% db)		
22.	*Iochroma gesnerioides* (leaves)	MAE: Microwave power = 25 W; Solvent = methanol; Solvent/feed ratio = 50; Extraction time = 40 s Soxhlet: Withaferin A (0.41% db a) 1. Solvent = water; Solvent/feed ratio = 6; Extraction time = 15 min 2. Solvent = methanol; Solvent/feed ratio = 100; Extraction time = 6 h	Withaferin A (0.48% db) Iochromolide (0.85% db) Withacnistin (0.39% db) Withaferin A (0.41% db) Iochromolide (0.81% db) Withacnistin (0.38% db)	Recovery of higher yields of bioactive compound compared to other extraction techniques.	[60]
23.	*Xanthoceras sorbifolia* Bunge. (yellow horn)	Microwave power = 900 W; Solvent = ethanol: water (40: 60 v/v); Solvent/feed ratio = 30; Temperature = 50 °C; Extraction time = 7 min × 3 cycles UAE: Microwave power = 250 W; Solvent = ethanol: water (40:60 v/v); Solvent/feed ratio = 30; Temperature = 50 °C; t = 60 min × 3 cycles Reflux: Microwave power = 800 W; Solvent = ethanol: water (40:60 v/v); Solvent/feed = 30; Temperature = 50 °C; Extraction time = 90 min × 3 cycles	Triterpene saponins (11.62% wb) Triterpene saponins (6.78% wb) Triterpene saponins (10.82% wb)	Recovery of higher yields of bioactive compound compared to other extraction techniques.	[48]
24.	*Ocimum basilicum* L. (basil)	SFME: Microwave power = 500 W; Temperature = 100 °C; Extraction time = 30 min	Eugenol (43.2% wb) Linalool (25.3% wb) Global yield (0.029% wb)	Recovery of higher yields of bioactive compound compared to other extraction techniques.	[61]

Number	Plant sample	Results obtained	Bioactive compounds	Remarks	Reference
		HD: Eugenol (11.0% wb a) s = water; S/F = 12; T = 100 C; t = 4.5 h	Eugenol (11.0% wb) Linalool (39.1%% wb) Global yield (0.028% wb)		
25.	*Mentha crispa* L. (gardenmint)	SFME: Microwave power = 500 W; Temperature = 100 °C; Extraction time = 30 min HD: Solvent = water; Solvent/feed ratio = 12; Temperature = 100 °C; Extraction time = 4.5 h	Limonene (9.7% wb) Carvone (64.9% wb) Global yield (0.095% wb) Limonene (20.2% wb) Carvone (52.3% wb) Global yield (0.095% wb)	Recovery of higher yields of bioactive compound compared to other extraction techniques.	[61]
26.	*Thymus vulgaris* L. (thyme)	SFME: Microwave power = 500 W; Temperature = 100 C; Extraction = 30 min HD: g -Terpinene (22.8% wb a) s = water; S/F = 12; T = 100 C; t = 4.5 h	γ-Terpinene (17.1% wb) Eugenol (51.0% wb) Global yield 0.160% wb) γ-Terpinene (22.8%% wb) Eugenol (40.5%% wb) Global yield 0.161% wb) Global yield (2.70% db)	Recovery of higher yields of bioactive compound compared to other extraction techniques.	[61]
27.	*Elletaria cardamomum* L. (cardamom)	SFME: Microwave power = 390 W; Temperature = 100 °C; Humidity = 67%; Extraction time = 75 min HD: Solvent = water; Solvent/feed ratio = 10; Temperature = 100 °C; Extraction time = 6 h	1,8-Cineole (26.23% db) Linalool (5.29% db) Terpin-4-ol (2.60% db) α-terpineol (3.88% db) Linalyl acetate (3.63% db) α-terpinyl acetate (45.45% db) 1,8-Cineole (26.23% db)	Recovery of higher yields of bioactive compound compared to other extraction techniques.	[62]
28.	*Gymnema sylvestre* R. Br.	MAE: Microwave power = 280 W; Solvent = methanol:water (85:15 v/v); Solvent/feed ratio = 25; Extraction time = 6 min Reflux:	Gymnemagenin (4.3% db) Gymnemagenin (3.3% db)	Recovery of higher yields of bioactive compound compared to other extraction techniques.	[63]

Number	Plant sample	Results obtained	Bioactive compounds	Remarks	Reference
		Solvent = methanol:water (85:15 v/v); Solvent/feed ratio = 100; T = 95 °C; Extraction time = 6 h			
		Maceration: Solvent = methanol:water (85:15 v/v); Solvent/feed ratio = 100; Extraction time = 24 h	Gymnemagenin (1.7% db)		
		Stirring: Solvent = methanol:water (85:15 v/v); Solvent/feed ratio = 100; Extraction time = 24 h	Gymnemagenin (2.2% db)		
29.	(*Melilotus officinalis* (L.) Pallas) (yellow sweet clover)	MAE: Microwave power = 100 W; Solvent = water: ethanol (50:50 v/v); Solvent/feed ratio = 20; Temperature = 50 °C; Extraction time = 5 min × 2 cycles	Coumarin (0.3978% db) *O*-coumaric acid (0.1257% db) Melilotic acid (0.9052% db)	Recovery of higher yields of bioactive compound compared to other extraction techniques.	[20]
		USAE: Solvent = water:ethanol (50:50 v/v); Solvent/feed ratio = 20; Extraction time = 60 min	Coumarin (0.3569% db) *O*-coumaric acid (0.1269% db) Melilotic acid (0.8092% db)		
		Soxhlet: Solvent = ethanol:water (95:5 v/v); Solvent/feed ratio = 16.67; Extraction time = 8 h	Coumarin (0.2156% db) *O*-coumaric acid (0.0708% db) Melilotic acid (0.6314% db)		
30.	*Salvia miltiorrhiza* Bunge. (dried root)	MAE: Solvent = ethanol:water (95:5 v/v); Solvent/feed ratio = 10; T = 80 °C; Extraction time = 2 min	Tanshinone IIA (0.29% db) Cryptotanshinone (0.23% db) Tanshinone I (0.11% db)	Recovery of higher yields of bioactive compound compare to other extraction techniques.	[64]
		Reflux: Solvent = ethanol:water (95:5 v/v); Solvent/feed ratio = 10; Extraction time = 45 min	Tanshinone IIA (0.25% db) Cryptotanshinone (0.24% db) Tanshinone I (0.11% db)		
		UAE: Solvent = ethanol:water (95:5 v/v); Solvent/feed ratio = 10; Extraction time = 75 min	Tanshinone IIA (0.28% db) Cryptotanshinone (0.25% db) Tanshinone I (0.10% db)		
		Soxhlet: Tanshinone IIA (0.33% db a) Solvent = ethanol:water (95:5 v/v); Solvent/feed ratio = 10; Extraction time = 95 min	Tanshinone IIA (0.33% db) Cryptotanshinone (0.25% db) Tanshinone I (0.33% db)		

Number	Plant sample	Results obtained	Bioactive compounds	Remarks	Reference
31.	*Radix astragali* (dried root)	MAE: Microwave power = 700 W; Solvent = ethanol: water (80:20 v/v); Solvent/feed ratio = 25; Temperature = 70 °C; Extraction time = 5 min × 3 cycles	Astragalosides I (0.788% db) Astragalosides II (0.351% db) Astragalosides III (0.206% db) Astragalosides IV (0.278% db)	Recovery of higher yields of bioactive compound compare to other extraction techniques.	[65]
		Soxhlet: Solvent = ethanol:water (80:20 v/v); Solvent/ feed ratio = 20; Temperature = 90 °C; Extraction time = 4 h	Astragalosides I (0.770% db) Astragalosides II (0.347% db) Astragalosides III (0.193% db) Astragalosides IV (0.242% db)		
		Reflux: Solvent = ethanol:water (80:20 v/v); Solvent/ feed ratio = 20; Temperature = 90 °C; Extraction time = 1 h	Astragalosides I (0.761% db) Astragalosides II (0.352% db) Astragalosides III (0.203% db) Astragalosides IV (0.257% db)		
		UAE: Power = 100 W; Solvent = ethanol:water (80:20 v/v); Solvent/feed ratio = 20; Extraction time = 40 min	Astragalosides I (0.519% db) Astragalosides II (0.302% db) Astragalosides III (0.190% db) Astragalosides IV (0.225% db)		
		Maceration: Solvent = ethanol:water (80:20 v/v); Solvent/ feed ratio = 20; Extraction time = 12 h	Astragalosides I (0.411% db) Astragalosides II (0.299% db) Astragalosides III (0.166% db) Astragalosides IV (0.206% db)		
32.	Sweet potato [*Ipomoea batatas* (L.) Lam.]	MAE: Microwave power = 123 W; Solvent = ethanol: water (53:47 v/v); Solvent/feed ratio = 25; Extraction time = 2 min	Total phenolics (6.115% db)	The recovery of phenolic compounds was higher in MAE compare to other techniques.	[66]
		CSE:	Total phenolics (5.969% db)		

Number	Plant sample	Results obtained	Bioactive compounds	Remarks	Reference
		Solvent = ethanol:water (60:40 v/v); Solvent/ feed ratio = 30; Extraction time = 120 min			
33.	Tobacco leaves	MAE: Microwave power = 700 W; Solvent = hexane: ethanol (1:3 v/v) + NaOH (0.05 mol/L); Solvent/feed ratio = 10; Extraction time = 40 min HRE: Solvent = hexane:ethanol (1:3 v/v) + NaOH (0.02 mol/L); Solvenf/feed ratio = 10; Temperature = 60 °C; Extraction time = 180 min	Solanesol (0.91% db) Solanesol (0.87% db)	The recovery of phenolic compounds was higher in MAE compare to other techniques.	[67]
34.	*Lavandula angustifolia* Mill., Lamiaceae (lavender flowers)	MSD: Microwave power = 200 W; flowrate = 8 g/ min; Extraction time = 6 min	1,8-Cineole (14.40% db) Linalool (42.52% db) Global yield (2.7% db)	Recovery of higher yields of bioactive compound	[68]
35.	*Radix astragali* (root of *Astragalus*; Huangqi)	MAE: Solvent = ethanol:water (95:5 v/v); Solvent/ feed ratio = 25; Temperature = 110 °C; Extraction time = 25 min × 2 cycles Soxhlet: Solvent = methanol; Solvent/feed ratio = 25; Temperature = 85 °C; Extraction time = 4 h UAE: Solvent = methanol; Solvent/feed ratio = 20; Temperature = 60 °C; Extraction time = 30 min × 2 cycles HRE: Solvent = ethanol:water (90:10% v/v); Solvent/ feed ratio = 25; Temperature = 75 °C; Extraction time = 2 h × 2 cycles	Flavonoids (0.1292%) Flavonoids (0.1190%) Flavonoids (0.0736%) Flavonoids (0.0934%)	Recovery of higher yields of bioactive compound compares to other extraction techniques.	[69]

Number	Plant sample	Results obtained	Bioactive compounds	Remarks	Reference
36.	Yellow onion	VMHG: Microwave power = 500 W; Pressure = 700 mbar; Power/feed = 1 W/g; Temperature = 81 °C; Extraction time = 26 min; Moisture content = 84.5% MHG: Pressure = 1 bar; Temperature = 100 °C; Extraction time = 23 min; Moisture content = 84.5% CSE: Solvent = methanol:water (80:20 v/v); Solvent/feed ratio = 10; Revolution = 8,000 rpm; Extraction time = 5 min	Quercetin (0.662% db) Global yield (3.18% db) Quercetin (0.283% db) Quercetin (0.890% db)	Recovery of higher yields of bioactive compound compares to other extraction techniques.	[70]
37.	Yellow soybeans (finely ground)	MAE: Microwave power = 600 W, Solvent = acetonitrile/water (2 mL, 80:20 v/v), sonicated with HCl 15 min, 1 min MAE. (sample, solvent, time)	Isoflavoids	Excellent efficiency and low consumption of solvent, sample, and time.	[71]
38.	Soybeans	MAE: Microwave power = 500 W, Temperature = 50 °C, Solvent = 25 mL of ethanol (50%), Extraction time = 20 min	Isoflavones (75%)	High reproducibility without degradation.	[72]
39.	Green tea leaves (*Camellia sinensis* L.)	MWE: Microwave power = 600 W, Temperature = 80 °C or 100 °C, Solvent = 120 mL of milli-Q water, Extraction time = 60 min	Flavanols	The yield of flavanol extract is higher compared to CWE, especially EGCG (*Epigallocatechin gallate*) concentration. More efficient at both 80 and 100 °C.	[73]
40.	Dried *Saussurea medusa* cells	MAE: Microwave power = 460 W, Solvent =10 mL of	Flavonoids (4.1%)	High selectivity compared to other extraction methods.	[74]

Number	Plant sample	Results obtained	Bioactive compounds	Remarks	Reference
		ethanol (80%), Extraction time = 6 min with 15 s power-on and 30s power-off			
41.	*Radix astragali*	MAE: Microwave power = 1,000 W, Temperature = 110 °C, Solvent = Ethanol (90%), Solvent/feed ratio = 25, Extraction time = 25 min	Flavonoids	The yield of flavonoids was closer to that of SOX with methanol and higher than that of UAE with methanol.	[32]
42.	*Platycladus orientalis* (book-leaf pine)	DMAE: Microwave power = 80 W, Solvent = 5 mL methanol (80%), Solvent/feed ratio = 500:1, Extraction time = 5 min	Flavonoids (1.72%)	Very short time and little solvent quantity required.	[75]
43.	*Saussurea medusa* Maxim	DMAE: Microwave power = 1200 W, Solvent = 2 L of ethanol (80%), Solvent/feed ratio = 50, Extraction time = 60 min	Flavonoids (4.97%)	In comparison with the same dynamic system without a microwave, it showed significant improvement.	[40]
44.	Longan peel	MAE: Microwave power = 500 W, Temperature = 80 °C, Solvent = 50 mL of ethanol (95%), Solvent/feed ratio = 10, Extraction time = 30 min	Total phenolic content (TPC = 96.78 mg/g), excellent scavenging ability comparing to synthetic antioxidant BHT	Very short time and little solvent quantity required.	[76]
45.	Plants of Labiatae, Verbenaceae, and Styracaceae	MAE: Microwave power = 750 W, Solvent = 20 ml of acetone (60%), Solvent/feed ratio = 20:1, Extraction time = 4 min	Total phenolic content (TPC = 23.8 mg GAE/g) in *Rosmarinus officinalis*	Higher yield in little time of extraction.	[16]
46.	Dried roots of *Rhodiola* L.	MAE: Microwave power = 400 W, Solvent = 5 mLof methanol (50%), Solvent/feed ratio = 5, Extraction time = 5 min	Salidroside and tyrosol (94.4–123%)	Good recoveries.	[77]

Number	Plant sample	Results obtained	Bioactive compounds	Remarks	Reference
47.	*Herba epimedii*	DMAE: Microwave power = 80 W, Solvent = ethanol (60%), Extraction time = 6 min	Flavonoids	The extraction yield of flavonoids obtained through DMAE was more compared to SOX, HRE, UE, and PMAE. Microwave saves time and generates lesser decomposition.	[78]
48.	Onion (*Allium cepa* L.)	MHG: Microwave power = 500 W, Extraction time = 23 min	Total phenolic content (58.29 mg GAE/DW) Yield (81.5%) Flavonol (41.9%)	Shorter extraction time.	[79]
49.	Red, yellow, white, and grelot onion (*Allium cepa*)	MHG: Microwave power = 500 W, Extraction time = 23 min	Flavonol	MHG remains the preferred method for the extraction of flavonoids compared to CSE.	[70]
50.	Sea buckthorn (*Hippophae rhamnoides*) by-product	MHG: Microwave power = 400 W, Extraction time = 23 min	Flavonoids	MHG showed much higher phenolic content with greater antioxidant activity in comparison to CSE.	[52]
51.	Onion by-product	VMHG: Pressure = 0.7 bar, Microwave power = 500 W, Extraction time = 26 min	Flavonoids	More antioxidants (total quercetin content) was extracted compared to MHG and CSE; an efficient procedure for extraction of heat-sensitive plant components.	[80]
52.	Olive leaves	MAE: Microwave power = 200 W, Solvent = 8 mL ethanol (80%), Solvent/feed ratio = 8:1, Extraction time = 8 min	Biophenols	The main compounds ranged from 631 (verbacoside) to 23,200 mg/kg (oleuropein).	[81]
53.	Grape skin and seeds	MAE: Pressure = 1–10 atm, Microwave power = 500 W, Temperature = 65–140 °C, Solvent = 20 mL of methanol (100%), Extraction time = 20 min	Phenolic compounds	Flavanols were mostly found in skin but absent in grape seeds; catechin was abundant in seeds.	[82]

Number	Plant sample	Results obtained	Bioactive compounds	Remarks	Reference
54.	Purple corn (*Zea mays* L.) cob	MAE: Microwave power = 555 W, Solvent = of 1.5 M HCl-ethanol (95%), Solvent/feed ratio = 20, Extraction time = 19 min	Anthocyanins (185.1 mg/100 g)	More efficient and rapid than CSE.	[83]
55.	Tomato paste	UMAE: Microwave power = 98 W, Frequency = 40 KHz of ultrasonic processing, Solvent/feed ratio = 10.6, Extraction time = 367 s	Lycopene (97.4%)	More efficient and rapid than UAE.	[84]
56.	Noni plant roots (*Morinda citrifolia*)	MAE: Microwave power = 720 W, Temperature = 60 °C, Solvent = 10 mL of ethanol (80%), Solvent/feed ratio = 100, Extraction time = 15 min	Anthraquinones (95.91%)	A higher yield has been obtained with higher antioxidant activity.	[54]
57.	Seeds, leaves, pulp, and fruits of sea buckthorn (*Hippophae rhamnoides*)	MAE: Microwave power = 150 W, Temperature = 60 °C, Solvent = 50 mL of ethanol, Solvent/feed ratio = 10, Extraction time = 20 min	Phenolic constituents (9.3–23.5 mg GAE/g) Rutin compound (365 mg/g)	Higher yields.	[85]
58.	Aloe (Liliaceae)	MAE: Microwave power = 340 W, Solvent = ethanol/water (20 mL, 80/20, v/v), Solvent/feed ratio = 15, Extraction time = 3 min	Aloe-emodin	Higher yield compared to other extraction methods.	[86]
59.	Bitter leaves	MAE: Microwave power = 558 W, Solvent = ethanol/water (76/24% v/v), Temperature = 70 °C, Solvent/feed ratio = 10, Extraction time = 4 min	Polyphenolic compounds	Several phenolic compounds were extracted.	[87]
60.	Purple fleabane	MAE: Microwave power = 444 W, Solvent = ethanol/water (47/53% v/v), Solvent/feed ratio = 14, Extraction time = 2 min	Total phenolic content = 85.64 $\pm$ 0.52 mg GAE/g d.w. Total flavonoid	Recovery of higher yield in a shorter time compared to Soxhlet extraction technique.	[21]

Number	Plant sample	Results obtained	Bioactive compounds	Remarks	Reference
			content = 52.72 ± 0.93 mg QE/g d.w.		
61.	Scent leaves	MAE: Microwave power = 553 W, Solvent = water, Solvent/feed ratio = 10, Extraction time = 3 min	Total phenolic content = 184.99 ± 3.05 mg GAE/g extract Total flavonoid content = 68.78 ± 2.07 mg QE/g extract	Recovery of higher yield in a shorter time.	[22]
62.	*Hibiscus sabdariffa*	MAE: Microwave power = 450 W, Solvent = ethanol/water (52/48%, v/v), Solvent/feed ratio = 15, Extraction time = 4 min	Total flavonoid content = 94.32 mg QE/g extract	Recovery of higher yield in a shorter time.	[6]
63.	*Chromolaena odorata* leaves	MAE: Microwave power = 493 W, Solvent = ethanol/water (51/49%, v/v), Extraction time = 3 min	Total phenolic content = 88.52 $mg_{GAEgDW}{}^{-1}$ Total flavonoid content = 68.84 $mg_{QEgDW}{}^{-1}$	Recovery of higher yield in a shorter time.	[9]

HRE: Heat reflux extraction, MHG: Microwave hydro-diffusion and gravity, MASD: Microwave-accelerated steam distillation, SD steam distillation, SFME: Solvent-free microwave extraction, MDG: Microwave dry-diffusion and gravity, MWHD: Microwave-assisted hydrodistillation; HD: Hydrodistillation, USAE: Soxhlet extraction, ultrasound-assisted extraction, MSD: Microwave steam distillation, VMHG: Vacuum microwave hydro-diffusion and gravity, CSE: Conventional solvent extraction.

Table 2.
Bioactive compounds extracted through MAE from different plant samples.

3.7 Influence of stirring

The influence of stirring can be linked to the mass transfer procedure in a solvent that causes convention. Hence, stability between vapor and aqueous phases can be obtained quickly. The process tends to accelerate through agitation, this enhances the dissolution and desorption of bioactive components in the plant sample [42]. Using a low solvent-to-feed ratio can be reduced as well as a reduction in the mass transfer barrier from solutes in a localized area emanating from inadequate solvent [26].

4. Previously extracted bioactive compounds from plants using MAE technique

MAE has been employed in several ways to extract bioactive compounds from different plant samples; the isolates from these plant samples are being used in nutraceutical and pharmaceutical applications. Microwave irradiation is mostly used to resolve some of the drawbacks associated with traditional methods. **Table 2** presents some of the previous studies that employed MAE to extractive bioactive compounds from plant samples. In the presented results obtained from previous studies as presented in **Table 2**, it can be seen the use of microwave-assisted extraction technique recover improved quantities of global yields, different phenolic compounds, and bioactive compounds. These indicated the efficacy of MAE over other methods of extractions.

5. Conclusions

This chapter outlines the studies and many advances in development in the MAE of a number of plant compounds. The factors that influence the performance of MAE technique have been extensively discussed as well as some of the bioactive compounds previously reported from plant samples using the MAE. The previously reported results showed that MAE can recover higher yields of bioactive compounds relative to other extraction methods. Thus, MAE is a promising method in achieving substantial bioactive compounds from plant materials due to its importance over other techniques.

Acknowledgements

Our gratitude goes to the Research and Innovation Department, Universiti Malaysia Pahang, Malaysia, for their support through the UIC-190806 Industrial research grant.

Conflict of interest

The authors declare that there is no conflict of interest.

Author details

Abdurahman Hamid Nour[1*], Alara Ruth Oluwaseun[1], Azhari Hamid Nour[2], Manal Suliman Omer[3] and Noormazlinah Ahmed[4]

1 Department of Chemical Engineering, College of Engineering, Universiti Malaysia Pahang-UMP, Malaysia

2 Faculty of Pure and Applied Sciences, International University of Africa, Khartoum, Sudan

3 Faculty of Industrial Management, Universiti Malaysia Pahang-UMP, Malaysia

4 Faculty of Chemical and Process Engineering Technology, Universiti Malaysia Pahang-UMP, Malaysia

*Address all correspondence to: nour2000_99@yahoo.com; abrahman@ump.edu.my

References

[1] S. B. Bagade, M. Patil, and S. B. Bagade, "Critical reviews in analytical chemistry recent advances in microwave assisted extraction of bioactive compounds from complex herbal samples: A review," *Crit. Rev. Anal. Chem.*, vol. 0, no. 0, pp. 1–12, 2019.

[2] M. Herrero, M. Plaza, A. Cifuentes, and E. Ibáñez, "Extraction techniques for the determination of phenolic compounds in food," *Compr. Sampl. Sample Prep.*, vol. 4, pp. 159–180, 2012.

[3] N. N. Azwanida, "A Review on the Extraction Methods Use in Medicinal Plants, Principle, Strength and Limitation," *Med. Aromat. Plants*, vol. 04, no. 03, pp. 3–8, 2015.

[4] K. A. Shams *et al.*, "Review Article Green technology: Economically and environmentally innovative methods for extraction of medicinal & aromatic plants (MAP) in Egypt," *J. Chem. Pharm. Res.*, vol. 7, no. 5, pp. 1050–1074, 2015.

[5] O. R. Alara, N. H. Abdurahman, and C. I. Ukaegbu, "Soxhlet extraction of phenolic compounds from *Vernonia cinerea* leaves and its antioxidant activity," *J. Appl. Res. Med. Aromat. Plants*, pp. 1–6, 2018.

[6] O. R. Alara, J. A. Alara, E. O. Obanijesu, and S. K. A. Mudalip, "Extract-rich in flavonoids from Hibiscus sabdariffa calyces: Optimizing microwave-assisted extraction method and characterization through LC-Q-TOF-MS analysis," *J. Food Process Eng.*, no. November, pp. 1–13, 2019.

[7] Z. Cai *et al.*, "Conventional, ultrasound-assisted, and accelerated-solvent extractions of anthocyanins from purple sweet potatoes," *Food Chem.*, vol. 197, p. Part A:266–272, 2016.

[8] O. R. Alara and N. H. Abdurahman, "Microwave-assisted extraction of phenolics from Hibiscus sabdariffa calyces: Kinetic modelling and process intensification," *Ind. Crops Prod.*, vol. 137, no. February, pp. 528–535, 2019.

[9] O. R. Alara, N. H. Abdurahman, and S. K. Adbul Mudalip, "Optimizing microwave-assisted extraction conditions to obtain Phenolic compounds-rich extract from Chromolaena odorata leaves," *Chem. Eng. Technol.*, 2019.

[10] R. N. Gedye *et al.*, "The use of microwave ovens for rapid organic synthesis," *Tetrahedron Lett.*, vol. 27, pp. 279–282, 1986.

[11] R. J. Giguerre, T. L. Bray, S. M. Duncan, and G. Majetich, "Application of commercial microwave ovens to organic synthesis," *Tetrahedron Lett.*, vol. 27, pp. 4945–4948, 1986.

[12] K. Ganzler, A. Salgo, and K. Valko, "Microwave extraction. A novel sample preparation method for chromatography," *J. Chromatogr. A*, vol. 371, pp. 299–306, 1986.

[13] P. Raut *et al.*, "Emerging Microwave Assisted Extraction (MAE) Techniques as an Innovative Green Technologies for the Effective Extraction of the Active Phytopharmaceuticals," *Res. J. Pharm. Technol.*, vol. 8, no. 6, pp. 655–666, 2015.

[14] F. Chemat and G. Cravotto, *Microwave-assisted extraction for bioactive compounds*. Springer, 2013.

[15] K. K. Chee, M. K. Wong, and H. K. Lee, "Microwave extraction of phthalate esters from marine sediment and soil," vol. 42, p. 378, 1996.

[16] C. Proestos and M. Komaitis, "Application of microwave-assisted extraction to the fast extraction of plant phenolic compounds," *LWT - Food Sci. Technol.*, vol. 41, p. 378, 2008.

[17] B. Kaufmann and P. Christen, "Recent extraction techniques for natural products: microwave-assisted extraction and pressurised solvent extraction," *Phytochem. Anal.*, vol. 13, pp. 105–113, 2002.

[18] X. Pan, G. Niu, and H. Liu, "Microwave-assisted extraction of tea polyphenols and tea caffeine from green tea leaves," *Chem. Eng. Process.*, vol. 42, p. 129, 2003.

[19] A. Longares-Patron and M. P. Canizares-Macias, "Focused microwaves-assisted extraction and simultaneous spectrophotometric determination of vanillin and p-hydroxybenzaldehyde from vanilla fragans," *Talanta*, vol. 69, p. 882, 2006.

[20] E. Martino, I. Ramaiola, M. Urbano, F. Bracco, and *S. collina*, "Microwave-assisted extraction of coumarin and related compounds from Melilotus of fi cinalis (L.) Pallas as an alternative to Soxhlet and ultrasound-assisted extraction," *J. Chromatogr. A*, vol. 1125, p. 147, 2006.

[21] O. R. Alara, N. H. Abdurahman, C. I. Ukaegbu, and N. H. Azhari, "*Vernonia cinerea* leaves as the source of phenolic compounds , antioxidants , and anti-diabetic activity using microwave-assisted extraction technique," *Ind. Crop. Prod.*, vol. 122, pp. 533–544, 2018.

[22] O. R. Alara, N. H. Abdurahman, C. I. Ukaegbu, and J. A. Alara, "Optimization of microwave-assisted extraction of phenolic compounds from *Ocimum gratissimum* leaves and its LC-ESI-MS/MS profiling, antioxidant and antibacteria activities," *J. Food Meas. Charact.*, 2020.

[23] S. S. Handa, S. P. S. Khanuja, G. Longo, and D. D. Rakesh, *EXTRACTION TECHNOLOGIES FOR MEDICINAL AND AROMATIC PLANTS*. 2008.

[24] L. Chen, D. Song, Y. Tian, L. Ding, A. Yu, and H. Zhang, "Application of on-line microwave sample-preparation techniques," *TrAC - Trends Anal. Chem.*, vol. 27, no. 2, pp. 151–159, 2008.

[25] G. Spigno and D. M. De Faveri, "Microwave-assisted extraction of tea phenols: A phenomenological study," *J. Food Eng.*, vol. 93, no. 2, pp. 210–217, 2009.

[26] C. H. Chan, R. Yusoff, G. C. Ngoh, and F. W. L. Kung, "Microwave-assisted extractions of active ingredients from plants," *J. Chromatogr. A*, vol. 1218, no. 37, pp. 6213–6225, 2011.

[27] Y. Tatke, P. and Jaiswal, "An overview of microwave assisted extraction and its applications in herbal reserach," *Res. J. Med. Plant*, vol. 5, no. 1, pp. 21–31, 2011.

[28] C.-H. Chan, J.-J. Lim, R. Yusoff, and G.-C. Ngoh, "A generalized energy-based kinetic model for microwave-assisted extraction of bioactive compounds from plants," *Sep. Purif. Technol.*, vol. 143, pp. 152–160, 2015.

[29] P. C. Veggi, J. Martinez, and M. A. a Meireles, *Microwave-assisted Extraction for Bioactive Compounds*. 2013.

[30] X. Pan, G. Niu, and H. Liu, "Microwave-assisted extraction of tea polyphenols and tea caffeine from green tea leaves," *Chem. Eng. Process.*, vol. 42, no. 2, pp. 129–133, 2003.

[31] C. S. Eskilsson and E. Björklund, "Analytical-scale microwave-assisted extraction," *J. Chromatogr. A*, vol. 902, pp. 227–250, 2000.

[32] W. Xiao, L. Han, and B. Shi, "Microwave-assisted extraction of flavonoids from Radix astragali," *Sep. Purif. Technol.*, vol. 62, no. 3, pp. 614–618, 2008.

[33] L. Chen, D. Song, Y. Tian, L. Ding, A. Yu, and H. Zhang, "Application of on-line microwave sample-preparation

techniques," *Trends Anal. Chem.*, vol. 27, pp. 151–159, 2008.

[34] F. Dahmoune, G. Spigno, K. Moussi, H. Remini, A. Cherbal, and K. Madani, "*Pistacia lentiscus* leaves as a source of phenolic compounds: Microwave-assisted extraction optimized and compared with ultrasound-assisted and conventional solvent extraction," *Ind. Crops Prod.*, vol. 61, pp. 31–40, 2014.

[35] B. G. Terigar, S. Balasubramanian, D. Boldor, Z. Xu, *M. lima*, and C. M. Sabliov, "Continuous microwave-assisted isoflavone extraction system: design and performance evaluation.," *Bioresour. Technol.*, vol. 101, no. 7, pp. 2466–71, 2010.

[36] C. W. Huie, "A review of modern sample-preparation techniques for the extraction and analysis of medicinal plants," *Anal. Bioanal. Chem.*, vol. 373, pp. 23–30, 2002.

[37] V. Mandal, Y. Mohan, and S. Hemalath, "Microwave assisted extraction-an innovative and promising extraction tool for medicinal plant research," *Pharmacogn. Rev.*, vol. 1, no. 1, pp. 7–18, 2007.

[38] L. Wang and C. L. Weller, "Recent advances in extraction of nutraceuticals from plants," *Trends Food Sci. Technol.*, vol. 17, pp. 300–312, 2006.

[39] Y. Li *et al.*, "Effect of energy density and citric acid concentration on anthocyanins yield and solution temperature of grape peel in microwave-Assisted extraction process," *J. Food Eng.*, vol. 109, no. 2, pp. 274–280, 2012.

[40] M. Gao, B. Z. Song, and C. Z. Liu, "Dynamic microwave-assisted extraction of flavonoids from Saussurea medusa Maxim cultured cells," *Biochem. Eng. J.*, vol. 32, no. 2, pp. 79–83, 2006.

[41] M. Desai, J. Parikh, and P. a. Parikh, "Extraction of Natural Products Using Microwaves as a Heat Source," *Sep. Purif. Rev.*, vol. 39, no. 1–2, pp. 1–32, 2010.

[42] G. H. Ruan and G. K. J. Li, "The study on the chromatographic fingerprint of Fructus xanthii by microwave assisted extraction coupled with GC-MS," *J. Chromatogr. B*, vol. 850, pp. 241–248, 2007.

[43] J.-Y. Hao, W. Han, S.-D. Huang, B.-Y. Xue, and X. Deng, "Microwave-assisted extraction of artemisinin from Artemisia annua L," *Sep. Purif. Technol.*, vol. 28, no. 3, pp. 191–196, 2002.

[44] D. Grigonis, P. R. Venskutonis, B. Sivik, M. Sandahl, and C. S. Eskilsson, "Comparison of different extraction techniques for isolation of antioxidants from sweet grass (Hierchloë odorata)," *J. Supercrit. Fluids*, vol. 33, pp. 223–233, 2005.

[45] X. Pan, H. Liu, G. Jia, and Y. Y. Shu, "Microwave-assisted extraction of glycyrrhizic acid from licorice root," *Biochem. Eng. J.*, vol. 5, pp. 173–177, 2000.

[46] H. Bagherian, F. Z. Ashtiani, A. Fouladitajar, and M. Mohtashamy, "Comparisons between conventional, microwave- and ultrasound-assisted methods for extraction of pectin from grapefruit," *Chem. Eng. Process*, vol. 50, pp. 1237–1243, 2011.

[47] Y. Chen, X. Ming-Yong, and G. Xiao-Feng, "Microwave-assisted extraction used for the isolation of total triterpenoid saponins from Ganoderma atrum," *J. Food Eng.*, vol. 81, pp. 162–170, 2007.

[48] J. Li *et al.*, "Optimization of microwaveassisted extraction of triterpene saponins from defatted residue of yellow horn (Xanthoceras sorbifolia Bunge.) kernel and evaluation of its antioxidant activity," *Innov. Food*

Sci. Emerg. Technol., vol. 11, pp. 637–664, 2010.

[49] P. S. Wakte, B. S. Sachin, A. A. Patil, D. M. Mohato, T. H. Band, and D. B. Shinde, "Optimization of microwave, ultrasonic and supercritical carbon dioxide assisted extraction techniques for curcumin from *Curcuma longa*," *Sep. Purif. Technol.*, vol. 79, pp. 50–55, 2011.

[50] S. Dhobi, M., Mandal, V., and Hemalatha, "Optimization of microwave assisted extraction of bioactive flavonolignan - silybinin," *J. Chem. Metrol.*, vol. 3, no. 1, pp. 13–23, 2009.

[51] M. Gallo, R. Ferracane, G. Graziani, A. Ritieni, and V. Fogliano, "Microwave assisted extraction of phenolic compounds from four different spices," *Molecules*, vol. 15, pp. 6365–6374, 2010.

[52] S. Périno-Issartier, S. Zill-e-Huma, M. Abert-Vian, and F. Chemat, "Solvent free microwave-assisted extraction of antioxidants from sea buckthorn (Hippophae rhamnoides) food by-products," *food Bioprocess Technol.*, vol. 4, pp. 1020–1028, 2010.

[53] S. Raghavan and M. P. Richards, "Comparison of solvent and microwave extracts of cranberry press cake on the inhibition of lipid oxidation in mechanically separated turkey," *Food Chem.*, vol. 102, pp. 818–826, 2007.

[54] S. Hemwimon, P. Pavasant, and A. Shotipruk, "Microwave-assisted extraction of antioxidative anthraquinones from roots of Morinda citrifolia," *Sep. Purif. Technol.*, vol. 54, pp. 44–50, 2007.

[55] G. Cravotto, L. Boffa, S. Mantegna, P. Perego, M. Avogadro, and P. Cint, "Improved extraction of vegetable oils under high-intensity ultrasound and/or microwaves," *Ultrason. Sonochem.*, vol. 15, no. 5, pp. 898–902, 2008.

[56] F. Chemat, M. E. Lucchesi, J. Smadja, L. Favretto, G. Colnaghi, and F. Visinoni, "Microwave accelerated steam distillation of essential oil from lavender: a rapid, clean and environmentally friendly approach," *Anal. Chim. Acta*, vol. 555, no. 1, pp. 157–160, 2006.

[57] A. Farhat, A.-S. Fabiano-Tixier, F. Visinoni, M. Romdhane, and F. Chemat, "A surprising method for green extraction of essential oil from dry spices: microwave dry-diffusion and gravity," *J. Chromatogr. A*, vol. 1217, no. 47, pp. 7345–7350, 2010.

[58] L. Hongyan, Z. Deng, T. Wu, R. Liu, S. Loewen, and R. Tsao, "Microwave-assisted extraction of phenolics with maximal antioxidant activities in tomatoes," *Food Chem.*, vol. 130, no. 4, pp. 928–936, 2012.

[59] Á. Kapás *et al.*, "The kinetic of essential oil separation from fennel by microwave assisted hydrodistillation (MWHD)," *UPB Sci. Bull. Ser. B*, vol. 73, no. 4, pp. 113–120, 2011.

[60] B. Kaufmann, P. Christen, and V. Jean-Luc, "Parameters affecting microwave-assisted extraction of withanolides," *Phytochem. Anal.*, vol. 12, no. 5, pp. 327–331, 2001.

[61] M. E. Lucchesi, F. Chemat, and J. Smadja, "Solvent-free microwave extraction of essential oil from aromatic herbs: comparison with conventional hydro-distillation," *J. Chromatogr. A*, vol. 1043, no. 2, pp. 323–327, 2004.

[62] M. E. Lucchesi, J. Smadja, S. Bradshaw, W. Louw, and F. Chemat, "Solvent free microwave extraction of Elletaria cardamomum L.: a multivariate study of a new technique for the extraction of essential oil.," *J. Food Eng.*, vol. 79, no. 3, pp. 1097–1086, 2007.

[63] V. Mandal, S. Dewanjee, and S. C. Mandal, "Microwave-assisted extraction

of total bioactive saponin fraction from Gymnema sylvestre with reference to gymnemagenin: a potential biomarker," *Phytochem. Anal.*, vol. 20, no. 6, pp. 491–497, 2009.

[64] X. Pan, G. Niu, and H. Liu, "Microwave-assisted extraction of tanshinones from Salvia miltiorrhiza Bunge. with analysis by high-performance liquid chromatography," *J. Chromatogr. A*, vol. 922, no. 1–2, pp. 371–375, 2001.

[65] M. M. Yan, W. Liu, Y. J. Fu, Y. G. Zu, C. Y. Chen, and M. Luo, "Optimisation of the microwave-assisted extraction process for four main astragalosides in Radix astragali," *Food Chem.*, vol. 119, no. 4, pp. 1663–1670, 2010.

[66] J. Song, D. Li, C. Liu, and Y. Zhang, "Optimized microwave-assisted extraction of total phenolics (TP) from Ipomoea batatas leaves and its antioxidant activity," *Innov. Food Sci. Emerg. Technol.*, vol. 12, pp. 282–287, 2011.

[67] H.-Y. Zhou and C.-Z. Liu, "Microwave-assisted extraction of solanesol from tobacco leaves," *J. Chromatogr. A*, vol. 1129, pp. 135–139, 2006.

[68] N. Sahraoui, M. A. Vian, I. Bornard, C. Boutekedjiret, and F. Chemat, "Improved microwave steam distillation apparatus for isolation of essential oils: comparison with conventional steam distillation," *J. Chromatogr. A*, vol. 1210, no. 2, pp. 229–233, 2008.

[69] W. Xiao, L. Han, and B. Shi, "Microwave-assisted extraction of fl avonoids from Radix astragali," *Sep. Purif. Technol.*, vol. 62, pp. 616–620, 2008.

[70] Zill-e-Huma, M. Abert-Vian, A. S. Fabiano-Tixier, M. Elmaataoui, O. Dangles, and F. Chemat, "A remarkable influence of microwave extraction: enhancement of antioxidant activity of extracted onion varieties," *Food Chem.*, vol. 127, pp. 1472–1480, 2011.

[71] M. Careri, C. Corradini, L. Elviri, and A. Mangia, "Optimization of a rapid microwave assisted extraction method for the liquid chromatography-electrospray-tandem mass spectrometry determination of isoflavonoid aglycones in soybeans," *J. Chromatogr. A*, vol. 1152, pp. 274–279, 2007.

[72] M. A. Rostagno, *M. Palma*, and C. G. Barroso, "Microwave-assisted extraction of soy isoflavones," *Anal. Chim. Acta*, vol. 588, pp. 274–282, 2007.

[73] E. Nkhili, V. Tomao, H. El Hajji, E. S. El Boustani, F. Chemat, and O. Dangles, "Microwave-assisted water extraction of green tea polyphenols," *Phytochem. Anal.*, vol. 20, pp. 408–415, 2009.

[74] M. Gao and C. Liu, "Comparison of techniques for extraction of flavonoids from cultured cells of Saussurea medusa Maxim," *World J. Microbiol. Biotechnol.*, vol. 21, pp. 1461–1463, 2005.

[75] L. Chen *et al.*, "Continuous determination of total flavonoids in Platycladus orientalis (L.) Franco by dynamic microwave-assisted extraction coupled with on-line derivatization and ultraviolet-visible detection," *Anal. Chim. Acta*, vol. 596, pp. 164–170, 2007.

[76] Y. Pan *et al.*, "Antioxidant activity of microwave-assisted extract of longan (*Dimocarpus longan* Lour.) peel," *Food Chem.*, vol. 106, pp. 1264–1270, 2008.

[77] Y. Mao, Y. Li, and N. Yao, "Simultaneous determination of salidroside and tyrosol in extracts of Rhodiola L. by microwave assisted extraction and high-performance liquid chromatography," *J. Pharm. Biomed. Anal.*, vol. 45, pp. 510–515, 2007.

[78] L. Chen *et al.*, "Dynamic microwave-assisted extraction of flavonoids from Herba epimedii," *Sep. Purif. Technol.*, vol. 59, pp. 50–57, 2008.

[79] Zill-e-Huma, M. Abert-Vian, J. F. Maingonnat, and F. Chemat, "Clean recovery of antioxidant flavonoids from onions: optimising solvent free microwave extraction method," *J. Chromatogr. A*, vol. 1216, pp. 7700–7707, 2009.

[80] Zill-e-Huma, M. Abert-Vian, M. Elmaataoui, and F. Chemat, "A novel idea in food extraction field: study of vacuum microwave hydrodiffusion technique for by-products extraction," *J. Food Eng.*, vol. 105, pp. 351–360, 2011.

[81] R. Japon-Lujan, J. M. Luque-Rodriguez, and M. D. Luque de Castro, "Multivariate optimisation of the microwave-assisted extraction of oleuropein and related biophenols from olive leaves," *Anal. Bioanal. Chem.*, vol. 385, pp. 753–759, 2006.

[82] A. Liazid, *M. Palma*, J. Brigui, and C. G. Barroso, "Investigation on phenolic compounds stability during microwave-assisted extraction," *J. Chromatogr. A*, vol. 1140, pp. 29–34, 2007.

[83] Z. Yang and W. Zhai, "Optimization of microwave-assisted extraction of anthocyanins from purple corn (Zea mays L.) cob and identi fi cation with HPLC-MS," *Innov. Food Sci. Emerg. Technol.*, vol. 11, pp. 470–476, 2010.

[84] L. Zhang and Z. Liu, "Optimization and comparison of ultrasound/microwave assisted extraction (UMAE) and ultrasonic assisted extraction (UAE) of lycopene from tomatoes," *Ultrason. Sonochem.*, vol. 15, pp. 731–737, 2008.

[85] U. K. Sharma, K. Sharma, N. Sharma, A. Sharma, H. P. Singh, and A. K. Sinha, "Microwave-assisted efficient extraction of different parts of Hippophae rhamnoides for the comparitive evaluation of antioxidant activity and quantification of its phenolic constituents by reverse-phase high-performance liquid chromatography (RP-HPLC)," *J. Agric. Food Chem.*, vol. 56, pp. 374–379, 2008.

[86] G. Wang, P. Su, F. Zhang, X. Hou, Y. Yang, and Z. Guo, "Comparison of microwave-assisted extraction of aloe-emodin in aloe with Soxhlet extraction and ultrasound-assisted extraction," *Sci. China Chem.*, vol. 54, no. 1, pp. 231–236, 2011.

[87] O. R. Alara, N. H. Abdurahman, C. I. Ukaegbu, Z. Hassan, and N. A. Kabbashi, "Dataset on LC-Q-TOF/MS tentative identification of phytochemicals in the extract of Vernonia amygdalina leaf through positive ionization," *Data Br.*, vol. 21, pp. 1686–1689, 2018.

Chapter 2

Microwave-Assisted Solid Extraction from Natural Matrices

Valeria Cavalloro, Emanuela Martino, Pasquale Linciano and Simona Collina

Abstract

The extraction of secondary metabolites from plants, and natural sources in general, is a cornerstone in medicinal chemistry and required the development of sustainable extraction techniques. Microwave-Assisted Solid Extraction (MASE) is a promising extractive methodology being more effective than traditional extraction techniques. It offers higher and faster extraction performance ability with less solvent consumption and protection toward thermolabile constituents. For these reasons, MASE resulted in a suitable extractive methodology in all aspects, including economical and practical, compared to traditional extraction techniques, especially over Soxhlet or solid–liquid extraction. In this chapter, a brief theoretical background about the use of microwave energy for extraction has been presented for better understanding. Then, the potential of MASE for the extraction of secondary metabolites from natural resources, for evaluating the plant productivity and for evaluating the quality of the natural matrices will be reviewed. The discussion is supported by reporting recent applicative examples of MASE applied to the extraction of the most representative chemical classes of secondary metabolites, with a special focus on some drugs or compounds of pharmaceutical and nutraceutical interest.

Keywords: MASE, plant material, natural matrices, secondary metabolite, bioactive compounds, plant productivity

1. Introduction

Nature Aided Drug Discovery (NADD) represents the most ancient approach in finding new active compounds for fighting human diseases, and still today it plays a crucial role in drug discovery [1]. New chemical entities (NCE) from natural derivation represent a relevant slice among the drugs approved by Food and Drug Administration (FDA) and the European Medicines Agencies (EMA) for commercialization and administration on humans [2]. More than half the total anti-infective drugs approved in the last forty years resulted from a NADD approach and a similar trend can be observed for anticancer drugs, where 41% of them derived from natural sources and only 16% are classifiable as totally synthetic small molecules [2]. Moreover, the Global Herbal Medicine Market Size is expected to increase to USD 129 billion by 2023, according to Market Research Future [3].

The success of NADD finds its main reason in the wider and heterogeneous chemical space covered by natural products whether compared with synthetic derivatives. The 83% of the chemical scaffolds found in natural compounds are unique and absent in synthetic NCE, due to the lack of commercially available synthons or cumbersome and prohibitive synthetic procedures [4]. Thus, the screening of libraries of compounds derived from natural sources still remains a worthy procedure for the identification of new and unexplored NCE. Besides, marine sources or lichens are still almost uninvestigated and might therefore represent an inestimable treasure of new potential drugs [5–7].

The natural compounds of interest in NADD are secondary metabolites that are not directly involved in the essential functions of the cell cycle and duplication processes and are characterized by high structural variability. From a structural standpoint, they are classified into alkaloids, terpenoids, saponins, lignans, flavonoids, and tannins [8]. Secondary metabolites are produced in different amounts for vexillary functions or defensive responses to biotic or abiotic stress being involved in the system of plant defense [9]. For this reason, specific secondary metabolites may be considered as markers of the plant health and may be used to evaluate the quality of the selected natural matrix and the effects that the environmental factors have on it. The evaluation of their content in natural sources, as well as their recovery, require the exploitation of ad hoc extractive procedures since secondary metabolites are mainly present in the intracellular domain. Thus, appropriate extraction procedures involve the rupture of the cell wall and cellular membranes, thus favoring the passage of the secondary metabolites into the extraction solvent, from which they can be recovered.

Over the years different protocols and techniques have been developed for the extraction of metabolites from natural matrices, named Solid–Liquid Extraction (SLEs). SLEs are classified in conventional (or traditional) and non-conventional (or innovative) methodologies [10]. In conventional methodologies, SLE is performed by heating the natural source with conventional heating sources (i.e., flame, heating plate, or mantle) in the presence or not of solvent and with or without stirring. Examples of conventional methodologies are maceration (digestion, infusion, and decoction), percolation, and Soxhlet or steam distillation. Solvents, or generally mixtures of solvents, with wide grades of polarity such as methanol, ethanol, acetone, ethyl acetate, trichloromethane, hexane, *etc*. are employed. Nevertheless, conventional extraction methods usually require a large volume of solvents, long extraction time, and high temperature. Such harsh extraction conditions may lead to thermal or chemical degradation of the metabolites, thus resulting in a low yield of the final extract. Moreover, the upscaling at an industrial level would be impracticable, owing to energy consumption, technological inaccessibility, and environmental considerations [11]. Conversely, non-conventional methodologies exploit innovative chemical–physical principles and cutting-edge technologies to facilitate the extractive procedures and the recovery of the product of interest from the natural source. Varied energy sources and extractive principles may be exploited and, therefore, specific equipments are needed. Microwave-assisted Solid Extraction (MASE), ultrasound-assisted extraction (UAE), pressurized solvent extraction (PLE), and supercritical fluid extraction (SFE) are the most frequently non-conventional methodologies extensively used for NADD [11].

In this chapter, the potential of microwave-assisted irradiation for the extraction of secondary metabolites from plants, and natural resources in general, will be discussed, with a special focus on recent applicative examples of the most representative chemical classes.

2. Microwave-Assisted Solid Extraction (MASE)

The physical principles underlying MASE are completely different from those of conventional extraction because microwave irradiation can cause a more effective disruption of the cellular structures (cell walls and cellular membranes) thus favoring the release of the cellular content and speeding up the extraction process. The interaction of microwave irradiation with the solvents, intracellular water, and ions causes an increase of the dipolar rotation and ionic conductivity of molecules with dipolar moments and ions, which results in a rise of the temperature inside the cell. The vaporization of the intracellular water causes the dehydration of the cellular wall and the reduction of its resistance. This, combined with the abrupt increment of the intracellular pressure, leads to the cell wall and membranes disruption, thus facilitating the passage of the secondary metabolites into the extractive solvent [12, 13].

This is the result of a synergistic combination of heat and mass gradient working in the same direction from the inside to the outside of the cell, as confirmed by Scanning Electron Microscopy (SEM) analysis (**Figure 1**) [14].

Monomodal and multimodal microwave systems also referred to as single- or multi-mode systems specifically designed for MASE are nowadays available. A single-mode system permits to focus the microwave radiation on a restricted zone where the sample is subjected to a much stronger electric field. Conversely, in a multimode system, the microwave radiation is randomly dispersed within the microwave cavity, where the sample is irradiated [15]. Both mono- and multi-modal microwave devices comprise a magnetron, which generates microwave radiation, a waveguide, which is used to propagate the microwave from the source to the microwave cavity, the applicator, where the sample is placed and a circulator that allows the microwave to move only in the forward direction. The main difference between the two systems relies on the applicator that in the case of a multimodal system is a closed cavity where the microwaves are randomly dispersed, whereas in the monomodal oven the extraction vessel is directly located in-line with the waveguide [15]. Both systems are effective in extracting metabolites from natural sources, and their use is usually related to the amount of natural matrix. Accordingly, the monomodal system is preferred when the amount of natural source to be treated is relatively low (milligram to gram scale), and vice versa. Microwave extraction may be performed using open or closed extraction vessel systems. The open vessel apparatus originated from a modification of domestic MW ovens. The extraction is conducted at ambient pressure and the vessel is directly connected to a condenser to avoid loss of solvent or volatile components. An evolution of the open vessel apparatus is the Focused Microwave-Assisted Soxhlet (FMASE) which combines the classical Soxhlet extraction (SE) technique with MW irradiation. In the closed vessel system the entire extraction vessel is located within the oven, thus allowing better control of the pressure and temperature during the extraction [16].

The closed-vessel system presents several advantages: i) a higher temperature than open vessel systems can be reached because of the increased pressure inside the vessel raises the boiling point of the solvents used, thus decreasing the time needed for the extraction; ii) the loss of volatile substances is completely avoided because they are confined within the vessel; iii) a low amount of solvent is required because no evaporation occurs and there is little or no risk of airborne contamination thus

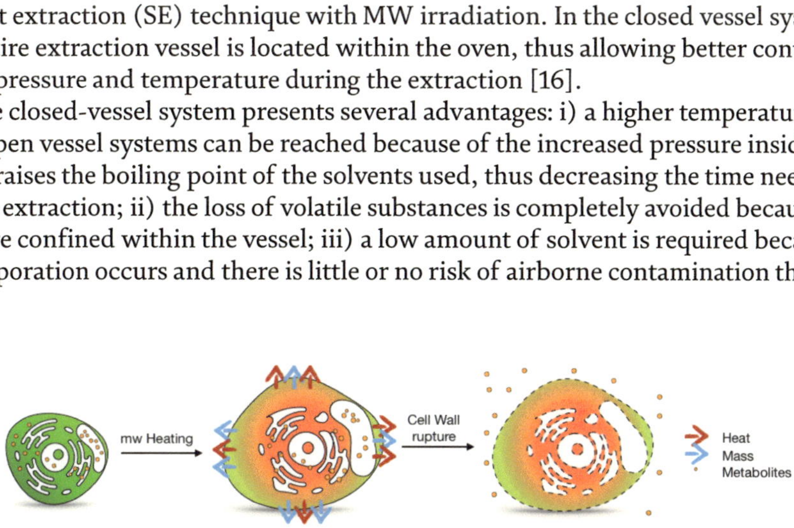

Figure 1.
Microwave heating effect on cells.

preventing the oxidation of air-sensitive metabolites; iv) the fumes are contained within the vessel, reducing the hazard of the extractive procedure.

In a closed vessel system high pressure can be reached. The high pressure stimulates various phenomena according to LeChatelier's principle, such as the transition of phase from one form to another, change in reaction dynamics, change in molecular structure, *etc*. hence resulting in the enhancement of extraction efficiency. Accordingly, working under high pressure causes alterations in the structure of some constituents of the cells such as lipids, proteins, enzymes, and outer cell membranes, thus damaging the plant wall and internal structure of the cell and reducing the mass transfer resistance. In this way, the secondary metabolites are released, leaving behind other cellular components. The High-Pressure MASE (HPMAE) is considered a newly emerging technique. It is a time-efficient, convenient, eco-friendly, safe, and energy-efficient extraction method when compared to the traditional or conventional methods of extraction [17].

Nevertheless, the use of closed-vessel systems presents some drawbacks: the amount of samples that can be processed is limited; the addition of reagents or solvents during operation is forbidden and the vessel must be cooled down before it can be opened to avoid loss of volatile constituents [16]. For processing a large amount of material, an open-vessel system is more appropriate. It allows the possibility to add reagents and to refill the solvent during the extraction, and to remove the excess of solvent during the extraction procedure. Moreover, the extraction of thermolabile metabolites is allowed since it usually reaches low temperatures relative to closed-vessel systems. On the other hand, the open-vessel systems ensure less reproducible results compared to the closed-vessel systems and the extraction in open-vessel could require a longer time to obtain results comparable to those achieved in closed-vessel [16].

The microwave energy may be applied also to extraction in solvent-free conditions (Solvent Free Microwave Extraction-SFME) [12]. In this case, the plant material is directly placed into the microwave reactor, without the addition of any solvent. The heating of the water contained by the plant material distends the natural matrices and causes the rupture of the cells releasing their content. This process has been successfully applied for the extraction of the essential oils. After MW heating, the volatile components co-evaporate with the in-situ water contained in the natural matrix; the vapors condense outside the microwave oven into a collector where they separate between essential oils and the aqueous phase. The latter is brought back into the vessel to refresh the amount of water in the sample, whereas the essential oil is collected apart [12].

3. Procedure set-up

The performance of the MASE process is strictly related to the operative conditions employed. Several parameters, such as solvent, solvent-drug ratio, temperature, time, pressure, microwave power, water content, and characteristic of the matrix must be optimized in setting up a MASE procedure. Each of these parameters should not be considered alone, but they are all linked together and the comprehension of the effects and influences of these factors is pivotal for MASE efficiency. A brief discussion about the role of these parameters in the design of a MASE protocol is herein reported.

3.1 Solvent

The selection of the solvent plays a crucial role in MASE, as well as in other conventional extraction processes, and several solvent parameters (solubility of the analyte, penetration, interaction with the matrix, dielectric constant, and mass transfer kinetic process) should be considered to perform the right choice. *In primis*, the solvent should assure the highest solubility of the analyte of interest while excluding undesired matrix components. However, conversely to the traditional extractive procedures, the chosen solvent must be able to absorb the microwave radiation and to convert it into heat. This depends on the dielectric constant and the dielectric loss of the solvent. Thus, in microwave application, solvents are usually classified in low, medium, or high absorbance whether they absorb at 300 W (i.e. carbon chlorides, 1,4-dioxane, tetrahydrofuran, diethyl ether, ethyl acetate, pyridine, toluene, chlorobenzene, xylenes, and hexane), 200 W (i.e. water, DMF, NMP, butanol, acetonitrile, HMPA, ketones, o-dichlorobenzene, 1,2-dichloroethane, 2-methoxy ethanol, acetic acid, and trifluoroacetic acid) or 100 W (i.e. DMSO, ethanol, methanol, propanol, nitrobenzene, formic acid, ethylene glycol), respectively. For low absorbing solvents, the heating rate can be increased by mixing with solvents with a higher dielectric constant or by adding salts to the mixture [18]. Recently, increasing interest in ionic liquids as MASE solvents or modifiers has been observed. Ionic liquids are organic salts with low melting points, and they are usually liquid at room temperature. They are characterized by extremely low vapor pressure, high stability and can solubilize both polar and non-polar metabolites. Moreover, they have the advantage to be eco-friendly solvents, although their extensive use is limited due to the high commercial costs. Another emerging eco-friendly alternative is the use of deep eutectic solvents. They possess physical and chemical properties similar to the ionic liquids, but they show better biodegradability, toxicity profiles, and solubility properties. Lastly, the content of water in the sample should be taken into account since it influences the heating rate and facilitates the transport of the analytes into the solvent at higher rates [19].

As stated in the previous paragraph, a Solvent-Free Microwave Extraction (SFME) is also possible. This procedure foresees the direct irradiation with MW of the plant material, fresh or rehydrated. A crucial role is played by the content of water in the sample, because it is the heating of such water to cause the rupture of the cells of the plant material, favoring the release of the content. This process is mainly applied for the extraction of volatile compounds such as essential oil. The oil evaporates by azeotropic distillation with the water contained in the sample. After cooling outside of the microwave reactor the oil separates from the water and can be collected through a modified Clevenger apparatus. Water is refluxed back into the microwave oven to allow the continuous extraction of the oil from the sample [12].

3.2 Liquid-Solid (L/S) ratio

Another important parameter to consider in the set-up of a MASE protocol is the ratio between the amount of sample and the volume of solvent. The latter should be enough to cover the sample during the entire process, especially when the matrix swells during the extraction process. Although in conventional extraction higher is the volume of solvent, higher is the yield of the extract, in MASE larger volume of solvent may result in more energy and time to heat the suspension and in a lower yield due to a non-uniform distribution and exposure to microwave. Usually, an L/S from 10:1 to 20:1 (mL/mg) is found to be the right ratio in many extractive processes reported in the literature [20].

Strictly correlated to the L/S ratio is the stirring rate since it affects the mass transfer process in MASE. However, the significance of this parameter is rarely explored. By stirring, the mass transfer barrier created by the concentrated compounds in a localized region due to insufficient solvent can also be minimized resulting in better extraction yield. In other words, agitation accelerates the extraction speed by accelerating the desorption and dissolution of compounds bound to the sample matrix [17].

3.3 Time

Extraction times in MASE lasts from a few minutes up to half an hour, and this represents an advantage for the extraction of thermal or oxygen labile compounds since it avoids the degradation of the compounds of interest [21]. The extraction yield is directly proportional to the extraction time, although it has been observed that this increment is very small for an extremely long time. Moreover, for longer extraction time overheating may occur, especially with high absorbent solvents, thus exposing thermolabile compounds to degradation. Whether longer extraction time is required, consecutive and shorter extraction cycles are preferable. The solvent can be collected after each extraction cycle and a fresh solvent could be added to the residue to guarantee the exhaustion of the matrix. This discontinuous procedure has been applied for the extraction of several secondary metabolites from plant material, allowing an enhanced yield and low decomposition of thermolabile compounds [22].

3.4 Temperature and microwave power

Temperature and MW power are strictly correlated. Power is the driving force of the process since it provides the energy necessary to excite the dipolar moments and the ionic conduction of the constituent of the sample, resulting in a proportional increase of the temperature and promoting the destruction of the natural matrix. Thus, the power of the microwave irradiation has to be carefully dosed in function of the amount of the sample, the solvent employed, the extraction time required, and the chemical stability of the secondary metabolites of interest [23]. Increasing the power results in an improved extraction yield and shorter extraction time. However, this result is true until the reaching of an optimal temperature beyond which a decrease in yield is observed, mainly due to the thermal stability of the target metabolite [24].

Accordingly, the temperature is a key parameter to enhance the efficacy of MASE and to avoid at the same time the degradation of the sample. Thus, the choice of the extraction temperature is strictly related to the properties of the solvent, the chemical stability of the metabolites of interest, and the microwave system used. In particular, at high temperatures the viscosity and the surface tension of the solvent diminish; moreover, the capability to solubilize the analytes, and to wet and penetrate the matrix increase, thus resulting in improved extractive efficacy. Also, when operating in a close-vessel, there is the advantage of heating the solvent above its boiling point, thus leading to a more performing extraction [25].

To conclude, the development of a proper MASE methodology must consider at least four variables: solvent, liquid/solid ratio, temperature, and time. To speed up the set-up of the procedure, the Design of Experiment (DoE) approach may be applied. This is a systematic statistic-based tool to assess the best experimental conditions both in the academic and industrial fields [26]. Thanks to this approach, all the variables and their interactions can be evaluated while doing the minimum number of runs.

Over the years, microwave-assisted extraction has been successfully applied to extract diverse classes of secondary metabolites (i.e. polyphenols, flavonoids, coumarins, terpenoids, cannabinoids, and alkaloids) from natural sources, for evaluating the plant productivity, for extracting bioactive compounds both for drug discovery or for commercial purposes.

Here below, studies of plant productivity based on MASE methodology and specific cases of extraction of natural compounds of pharmaceutical and nutraceutical interest will be discussed, with a special focus on resveratrol, terpenoids, and cannabinoids.

4. MASE procedures successfully applied to secondary metabolites extraction

4.1 Evaluation of plant productivity

Numerous applications report about the use of the MW to assist the extraction of organic and organometallic compounds from various matrices (soils, sediments, water samples, botanicals), with special emphasis on environmental applications [27, 28]. Extraction of natural matrices is essential to compare their productivity under different stress conditions [9], harvesting time [29], and places [30]. MASE offers the possibility of performing multiple extractions and therefore, it is suitable for the rapid screening of a numerous set of samples to evaluate the productivity of organisms.

An example is a work performed by Martino et al., regarding the MASE of *Melilotus officinalis,* harvested in different environmental situations, to compare the amount of coumarin and related compounds, and to find the best condition for its cultivation [22]. The Authors developed a rapid, reliable, and reproducible method of extraction from *M. officinalis* inflorescence of coumarin (**Figure 2**), melilotic acid, and o-coumaric acid, considered as productivity markers of the plants. A comparison of different extraction techniques evidenced that MASE is the best procedure in terms of both yields and extraction time [22]. The optimal extraction conditions consisted of two successive irradiations of 5 min each at 50°C, with a cooling step in between, and it resulted suitable for application to large sets of samples [22].

Another example is the setup of a fast and reproducible extraction methodology of vitexin and its isomer isovitexin from *Crataegus monogyna* (**Figure 2**) for evaluating the plant productivity and determining the best ecological conditions for hawthorn cultivation in northern Italy (Lombardy). These metabolites have a high pharmaceutical value due to their anti-hyperalgesic and neuroprotective effects and their activity against oxidative stress, cancer, and inflammation [31].

Within this context, Martino et al. set up a MASE procedure that can be applied for quantitative extraction of both metabolites from *C. monogyna* in just 3 minutes [50% aqueous methanol (v/v), 120°C, 120 W], bringing advantages both in terms of time (3 min *vs* 6 hours) and solvent consumption (0.05 vs 0.10 g/mL) over standard extraction methods [30]. The developed MASE protocol combined with isocratic

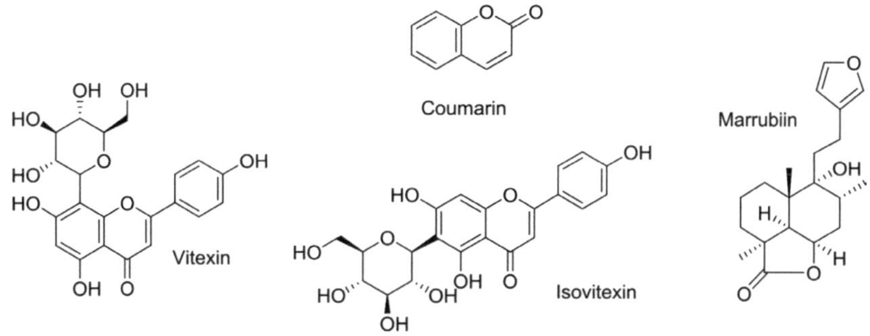

Figure 2.
Secondary metabolites extracted via MASE approach and considered as markers of the plant productivity.

HPLC analysis is suitable for the rapid screening of plant materials collected in different environmental conditions, and to determine the best ecological conditions for its cultivation. To extract vitexin and isovitexin from *Crotalaria sessiliflora,* Tang et al. exploited a microwave-assisted cloud-point extraction (MACPE). MACPE combines cloud-point extraction (CPE) with MAE. This has emerged as a technique to extract and separate bioactive compounds from medicinal plants [32]. Of note, using MACPE, hydrophobic compounds present in the aqueous phase can be favorably extracted into the hydrophobic core of micelles [33]. Applying MACPE, vitexin and isovitexin have been obtained in high yields and short times [34].

MASE can also be applied to study the effect of micronutrients or pollutants on secondary metabolites production. Amri et al. investigated the impact of soil copper (II) concentrations on nutrient uptake and the antioxidant system of *Marrubium vulgare*. Owing to waste deposition and agricultural practices, copper (II) tends to accumulate in high and toxic concentrations, leading to an alteration of the vital physiological or biochemical functions of the plants. As it is the case of *M. vulgare,* these effects may have a great impact on human health, since such a plant is used worldwide for its medicinal properties. To perform the study, the Authors selected marrubiin (**Figure 2**) as a reference compound, since it is the main secondary metabolite produced by *M. vulgare* leaves. A MASE protocol was developed for the easy extraction of marrubiin. This procedure allowed to evaluate the quality of a wide range of samples of white horehound. To optimize the process, the Authors used the statistical DoE approach. DoE findings indicated that the highest extraction efficiency of marrubiin with high repeatability was obtained using 100% ethanol at 120°C for 15 min, with significant benefits in terms of extraction times and environmental impact, given that ethanol is completely biodegradable. The MASE methodology may be applied for the characterization of *M. vulgare* herbal drug samples, thus evaluating their exposure to abiotic stress, revealing their phytochemical status, and facilitating the identification of raw materials obtained from a plant grown under stress conditions.

To sum up, MASE procedures is a versatile technique suitable for the evaluation of the plant productivity, and to assess the quality of vegetal matrices, since it is fast, reproducible, suitable for extraction of a large number of samples and requires a low amount of natural matrix.

4.2 Extraction of secondary metabolites for drug discovery or commercial purposes

4.2.1 Alkaloids

Alkaloids are a well-known class of secondary metabolites characterized by basic nitrogen. Over the years, many active alkaloids have been extracted via MW irradiation, e.g. ephedrine alkaloids, cocaine, and ergot alkaloids [35–37]. Unfortunately, results obtained for many of them have been comparable or worst if compared with the traditional method [38]. Nevertheless, microwaves have also spurred the discovery of new active alkaloids at the early stage of drug discovery. MASE protocols can be exploited to extract different alkaloids (examples are reported in **Figure 3**) from different botanicals like tuberous roots, leaves, and seeds [37, 39, 40].

As an example of MASE applied to the extraction of alkaloids, Pan et al. obtained a good recovery of caffeine and polyphenols from the leaves of green tea (*Thea sinensis* L.). MASE provided high extraction and selectivity, required a short time, and less labour-intensive, thus resulting in an efficient method in comparison with the conventional extraction procedures [41].

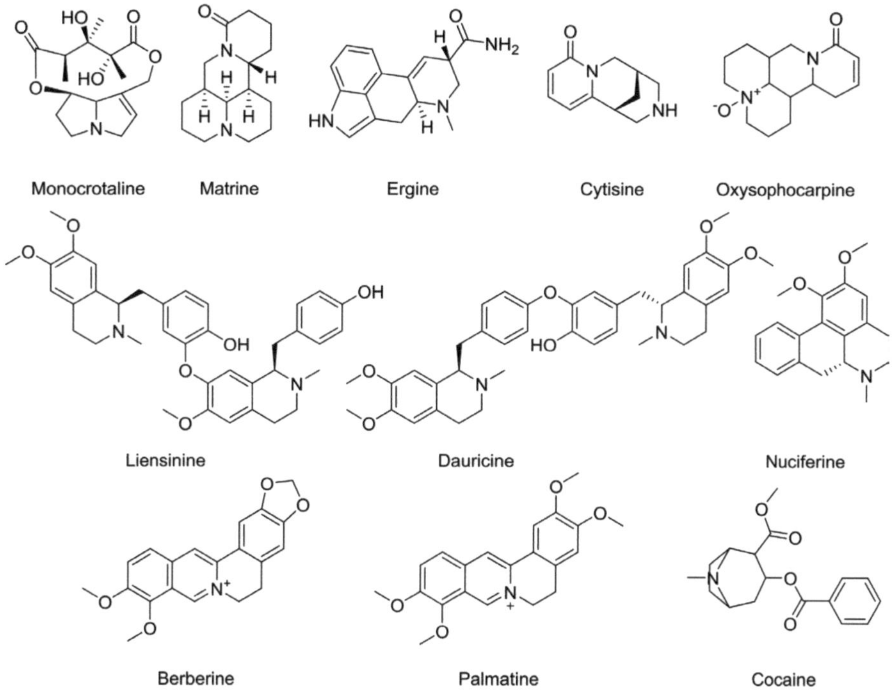

Figure 3.
Examples of alkaloids extracted via the MASE approaches.

Xiong et al. developed an efficient MASE protocol, within a drug discovery process, for the isolation of bioactive alkaloids (e.g. liensinine, isoliensinine, neferine, dauricine, nuciferine, **Figure 3**) from *Lotus plumule*. The optimal extraction conditions required a 65% aqueous methanol as a solvent and irradiation at 200 W for 260 seconds [42]. Another interesting example, reported by Zhou et al., is the microwave-assisted aqueous two-phase extraction, useful for rapid and simultaneous extraction and separation of alkaloids. This technique was applied to *Radix Sophorae tonkinensis*. The optimum conditions were summarized as follows: ethanol/Na_2HPO_4 as the extraction solvent, 100 mesh as particle size, 1:75 of S/L ratio, irradiating at 90°C for 5 min [43]. Matrine, sophocarpine, oxymatrine, oxysophocarpine, 5α-hydroxysophocarpine, sophoranol, cytisine, N-methylcytisine, and sophoridine were efficiently extracted.

Recently, Belwal et al. reported an optimized MASE protocol, defined by multicomponent analysis, for the extraction of berberine (**Figure 3**) and polyphenols from diverse species of *Berberis*. The medical properties of berberine (anti-diabetic, hepato-protectant, anti-arthritic, antioxidants, anti-microbial, neuro-protective, and hypo-lipidemic activity) are widely recognized, and it is used in pharmaceuticals and nutraceuticals preparation. In this study, multi-component analysis (MCA) has been used to extract berberine and polyphenols from *B. jaeschkeana* roots under microwave-assisted extraction (MAE) conditions. All the variables, above described, were considered under 42 experiments and the results of the model showed significant model fitness. Under optimized MAE condition, (i.e. 100% methanol, pH 2.0, 598 W, 2 min of irradiation time), the berberine and palmatine (**Figure 3**) contents were recorded in 4.6% and 2.0%, respectively. Under the optimized condition, the yield of alkaloids was found closer to the models' predicted value [34].

Regarding the alkaloids employed as drugs, or of interest for the toxicological use and/or abuse, few extractive procedures by MASE are reported in the literature. As an example, Brachet et al. extracted cocaine and benzoylecgonine from the leaves of *Erythroxylum coca* by MASE. Different solvents, particle size, time, and power were evaluated. Since MeOH is a high absorbing microwave solvent, and cocaine is highly soluble in it, it was found to be the best extraction solvent [36]. Interestingly, MASE found application in the forensic field as a rapid and cleanup-free method for the extraction and quantification of drugs of abuse and the respective metabolites from human fluids and tissues. Fernandez et al. reported the simultaneous extraction of cocaine, benzoylecgonine, cocaethylene, morphine, 6-monoacethylmorphine, and codeine from human urine [44], hair [45], and vitreous humor samples [46]. The MASE procedure reduces the extraction time, avoids the cleanup steps, and allows a quantitative recovery of the drugs.

4.2.2 Stilbene-based polyphenolic compounds

Stilbene-based polyphenolic compounds, i.e. resveratrol, pterostilbene, and piceatannol, are of particular interest from a medicinal chemistry standpoint, having multiple pharmacological activities (**Figure 4**).

In particular, *trans*-resveratrol (3, 5, 4′-trihydroxystilbene) became popular as a result of an attempt to explain the "French paradox" [47]. Resveratrol and other polyphenolic-stilbene derivatives showed a wide range of beneficial physiological properties. They possess antibacterial, anti-inflammatory, hypolipidemic, cardiovascular-hepatoprotective, and anticancer activities [48–50]. In particular, the hypolipidemic and cardiovascular protective activity derived from the agonistic activity against PPARα and PPARγ receptors [51, 52] For all this benefit, resveratrol has attracted the attention of the scientific community and pharmaceutical and nutraceutical industries. Indeed, several drugs and dietary supplements containing resveratrol are commercially available.

Even though resveratrol is produced naturally in plants, the extraction of resveratrol in commercial quantities is a problem, because of its low concentration, multiple steps of isolation and purification, unfriendly environmental issues, and seasonal occurrence [53]. Moreover, the preparation of resveratrol by synthesis is difficult owing to the formation of many unwanted side products [54, 55]. Only recently, the production of resveratrol in heterologous engineered microorganisms

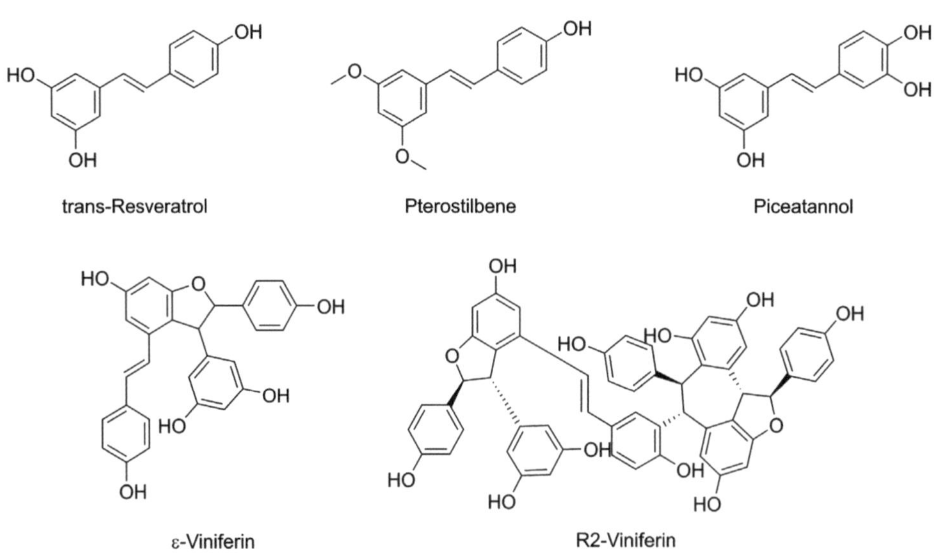

Figure 4.
Chemical structure of polyphenolic-stilbene based secondary metabolites.

has been proposed [56]. Thus, resveratrol is still being extracted from wild *Polygonum cuspidatum*'s root (Japanese knotweed), grape skins and seed, and the domestic giant knotweed of China, which is the world's largest producer [53, 56–58]. Garcia-Ayuso *et al.,* in late 1998, found that by applying the MW irradiation to SE, the last of the extraction was drastically reduced from 8 hours to 60 minutes with comparable results to SE in yield. MASE was further optimized by testing solvents and times on bark extraction and compared to SE on the same tree sample. The results suggested that microwave extraction may be more efficient than SE.

The extraction of resveratrol by MASE from different plant materials (i.e. *Arachis repens* and grape seeds) has also been investigated. To exhaustively extract resveratrol from *A. repens*, commonly known as peanut grass, three different methodologies (conventional maceration, ultrasound-assisted extractions, and MASE) have been compared. Although sonication resulted more effectively in the extraction of resveratrol compared to MASE and maceration, MASE showed to be an excellent choice since it extracted high yields in a reduced time [59].

In another study, Dang et al. combined the aqueous two-phase extraction technique (ATPE) with MASE for the extraction of the total polyphenol content, including resveratrol, from grape seeds [60]. Microwave-assisted ATPE (MAATPE) required lower solvent concentration and less time compared with other methods such as refluxing solvent or SE. A higher level of resveratrol was obtained with MAATPE, in contrast to ATPE. The Authors also compared the effectiveness of three solvents (water, water: ethanol (1:1) and ethanol) and three extraction methods, including MASE and ultrasound-assisted extraction (UAE) and the conventional SE. MASE provided a better extraction with water and ethanol (1:1) obtaining extracts very rich in polyphenolic substances, including stilbenes.

Lastly, MASE has successfully applied also for the extraction of other polyphenolic-stilbene based compounds such as pterostilbene, mainly found in blueberries and in *Pterocarpus marsupium* heartwood, and ε-viniferin, found in *Vitis coignetiae*, a wild grapevine (**Figure 4**). Kim et al. reported the MASE of pterostilbene, and other derivatives, from *Vitis coignetiae*, using 80% ethanol at 90 W for 15 min, resulting in a stilbenoids overall yield of 0.13%, with pterostilbene the most representative compound in the extract [61]. An optimized protocol (70–150 W for 8–18 min, using 30–50% ethanol) was further developed for the extraction of viniferin from the same drug [48]. Recently, Pinero et al. disclosed a new process for recovering stilbenes from woody vine by-products such as grape stem and cane samples. MASE was carried out under different extraction conditions. The best results were achieved from grape stems, using 80% ethanol in water as an extraction solvent, a temperature of 125°C, an irradiation power of 750 W for 5 min [49].

4.2.3 Terpenoids

4.2.3.1 Artemisinin and paclitaxel

Terpenes and isoprenoids, in general, gained much attention for their physiological functions (i.e., hormones, aliphatic membrane anchors, maintaining membrane structure), ecological roles (i.e., defense compounds, insect/animal attractants), and extensive pharmaceutical applications such as flavors, fragrances, and medicines.

In particular, artemisinin and paclitaxel represented two milestones in the fight against malaria and cancer, respectively. Artemisinin (**Figure 5**) is a sesquiterpene lactone isolated from *Artemisia annua* and it is a first-class drug for the treatment of drug-resistant malaria. The conventional artemisinin extraction procedure requires room temperature, heat-reflux, or SE. Hao et al., in 2002 reported a first attempt to extract artemisinin from *Artemisia annua* by MASE. Several solvents were explored,

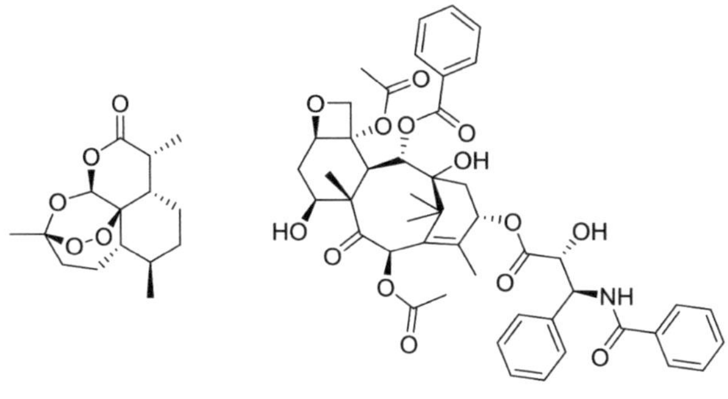

Figure 5.
Chemical structures of artemisinin and paclitaxel.

such as ethanol, trichloromethane, cyclohexane, n-hexane, petroleum ether, and two in-house oils. Compared with SE, supercritical CO_2 extraction, and normal stirring extraction, MASE of artemisinin from *Artemisia annua* considerably reduced the processing time to 12 minutes and resulted in a 92.1% extraction rate (compared to several hours and 60% extraction yield obtained with Soxhlet) [50]. Later, Liu et al. applied the MASE for the isolation and quantification of artemisinin in comparison with the traditional protocols. MASE confirmed shorter extraction time necessity, reduced solvent consumption, and higher recovery of artemisinin than conventional procedures. The best extraction solvent was petroleum ether–acetone (4:1 v/v), because of the high solubility of artemisinin and adequate microwave energy absorption, at 50°C. The highest yield of artemisinin achieved was 0.55% in 30 minutes among all the extractive methods used [62].

Recently, Misra et al. developed a rapid and reliable MASE and HPTLC protocol for the analysis of artemisinin. The optimized MASE conditions required 100 mg of dried and grinded drug with a size of 14 mesh dispersed into 10 mL of toluene. The irradiation of the sample at 160 W for 120 seconds led to the extraction of 0.816% of the content of artemisinin.

Paclitaxel (**Figure 5**) is a member of the taxane class, and it is one of the most important anticancer drugs approved for human use against ovarian, breast, and pulmonary cancer.

Although the total synthesis of paclitaxel has been reported, its application for the commercial production of this drug is impracticable. Thus, paclitaxel is still produced by extraction from taxol biomass. The most commonly used methods for the extraction of paclitaxel require the use of methanol at ambient temperature, although other protocols requiring refluxing methanol, 1:1 methanol-chloroform at ambient temperature, and percolation using ethanol or 95% ethanol-water at ambient temperature have been reported. However, these methods require a long time (12–24 h) for a complete extraction. Incorvia-Mattina et al. reported for the first time in 1997 the use of MASE to optimize the efficiency of the extraction of paclitaxel. The effects of the biomass, solvent ratio and water content on taxane recovery were also determined. Under appropriate MASE conditions an extract equivalent to the one obtained by conventional extraction methods was produced [63].

Talebi et al. investigated the use of MASE to extract paclitaxel from the needles of *Taxus baccata L.* The extraction parameters were investigated resulting in 90% aq. MeOH as a solvent, a temperature of 95°C, 7 min of extraction time, and a closed-vessel system as the best performing extractive conditions [64].

Recently, another study for the extraction of paclitaxel from biomass through MASE and based on kinetic and thermodynamic analysis has been carried out. The majority of paclitaxel was recovered from the biomass (~99%) within 6 min in a single cycle of microwave-assisted extraction at microwave powers of 50–150 W and temperatures of 30–45°C [64].

4.2.4 Phytocannabinoids

Cannabis sativa L. has always been considered a controversial plant due to its use as both medicine and illicit drug. Nevertheless, Cannabis is a good source of nutrients, fibers, and natural compounds thus, its industrial and pharmaceutical use is undoubtful. Cannabis produces a peculiar class of natural compounds, namely phytocannabinoids. The two most important and renowned phytocannabinoids are the cannabidiol derivatives (i.e. CBD, CBDV, CBDB, and CBDP) [65, 66] and the tetrahydrocannabinol derivatives (i.e. Δ^9-THC, Δ^9-THCV, Δ^9-THCB, and Δ^9-THCP) [66, 67] reported in **Figure 6**. Δ^9-THC is responsible for the recreational use of hemp and therefore its use is banished or tightly regulated by national governments.

CBD-like derivatives are non-psychotropic compounds but with other recognized pharmacological properties such as anti-inflammatory, antioxidant, and anticonvulsant. As an example, Epidiolex, a CBD-based anticonvulsant drug, has been approved in 2018 by Food and Drug Administration for the treatment of seizures associated with Lennox–Gastaut syndrome (LGS), Dravet syndrome, or tuberous sclerosis complex (TSC) in patients 1 year of age and older.

The discovery of a plethora of pharmacological activities ascribed to CBD and other minor phytocannabinoids has increased attention from both scientists and industries for medical, nutraceutical, and cosmetic applications of these cannabinoids.

Several synthetic procedures have been developed and optimized for the industrial preparations of phytocannabinoids and in particular of CBD. However, this process suffers from several drawbacks such as the cost of the starting materials, reagents and solvents, the formation of by-products with consequent cumbersome purification procedures, and the difficulty to control the stereochemistry, the isomerism of the terpenic double bond, and the easy interconversion of CDB into THCs in the synthetic conditions.

Thus, the extraction and purification of phytocannabinoids from *C. sativa* remain the preferred procedure for its cost-effectiveness. Besides, tight monitoring of the chemical consistency of the extracts results therefore mandatory in producing consistent and reliable medical cannabis preparations for human uses. Recently, Nahar et al. reviewed all the procedures adopted at the present for the extraction of naturally occurring phytocannabinoids [68].

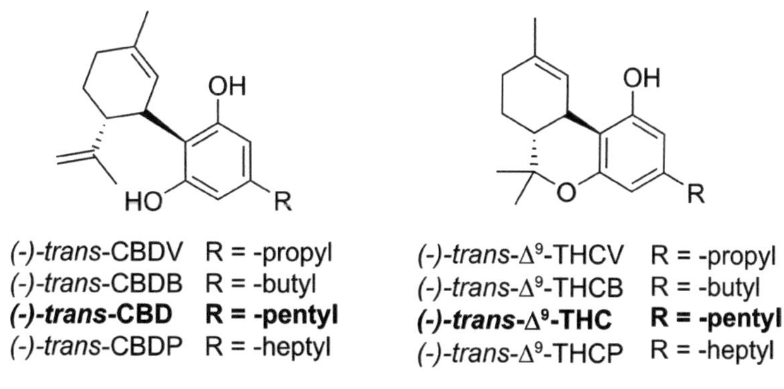

Figure 6.
Chemical structures of CBD-like and THC-like major phytocannabinoids present in C. sativa.

Focusing on MASE, Lewis-Bakekr et al. investigated the potential to directly extract and decarboxylate dried Cannabis material with the microwave reactor [69]. Dried plant material, suspended in ethanol, was subjected to heating with stirring in a microwave reactor at 150° C. Extraction yield for the concentrated resin was in the range of 19.6–24.4% and it resulted directly proportional to the heating time and dependent on the cultivar employed in the process. Interestingly, a complete decarboxylation of the phytocannabinoids was achieved in one step following this process and no acid forms of phytocannabinoids such as Δ^9-THCA and CBDA were detected in the resulting extract. Thus, MASE proves to be a worthy method for extraction and decarboxylation of phytocannabinoids due to the possibility to apply controlled temperatures and shorter extraction times. Moreover, this procedure ensures a more consistent and reproducible *Cannabis* extract with consequent reproducible efficacy of the therapeutic results. Kore et al. investigated and optimized the MASE process applied to *C. sativa* resulting in a patent application where they disclosed an improved method for extracting and decarboxylating cannabinoids from cannabis plant material, before, during, or after extraction [70]. MASE was compared to or used in tandem with other extraction strategies such as ultrasound extraction, SE, and supercritical fluid extraction.

The effect of time and temperature was investigated first. Extraction and decarboxylation of phytocannabinoid resulted in time and temperature dependence. To obtain 100% decarboxylation, the temperature must be sustained over a period without the burning of the cannabis material or the boiling/evaporation of the solvent. Because the solvent of choice is ethanol (b.p. 78°C at 1 atm), to reach a higher boiling temperature (i.e. 100–170°C) the extraction process must be carried out in a sealed vessel and under pressure. 170°C was the highest operative temperature achieved since higher temperatures (>180°C) resulted in the microwave run abortion due to the high pressure reached within the vial.

The extraction of cannabis by MASE at 100°C, 130°C, 150°C, and 170°C for 10 minutes resulted in a 23–25% yield of extract.

Interestingly, it appeared that the addition of a second step, such as SFE, after the MASE did not change the cannabinoid profile in the extract. Thus, MASE alone can perform an almost complete extraction of the cannabinoids from the cannabis plant material. Besides, the extraction and conversion of THCA and CBDA into THC and CBD was better at a temperature above 130°C, than at 100°C.

MASE was compared with the effectiveness of the commonly employed extractive procedure, namely maceration in ethanol, SE, and SFE. The conventional extractive procedures resulted in a low concentration of Δ^9-THC, THCA, and CBD, whereas the addition of the microwave step resulted in a significant increase in the concentration of CBD and THC. As expected, no THCA was detected.

To sum up, a worth general procedure for the extraction and decarboxylation of CBD and THC from cannabis plant material can be thus resumed: i) the drug is weighed and macerated in a mortar; ii) the grinded drug is charged in a microwave vial along with a stir bar; iii) the drug is submerged with ethanol and the vial is sealed; iv) the vial is irradiated with MW using the following conditions [a) Pre-stirring = 30 sec; b) run time = 10 min; c) temperature = 150°C; d) absorption = Normal]; v) the suspension is filtered, and the filtrate concentrated; iv) residual plant material may be subjected (but not necessarily) to SFE.

Drinic et al. extended these studies over other polyphenols and flavonoids as well as phytocannabinoids [71]. In particular, the effects of different extraction parameters, namely ethanol concentration, extraction time and solid/liquid ratio on extraction yield, total phenol content, total flavonoid content, antioxidant activity, reductive capacity, CBD content, and THC content were investigated. For MASE, a domestic microwave oven and a round-bottom flask connected with a condenser were used. The solid drug was mixed with the solvent (30, 50, or 70% v/v ethanol) in the selected solid/liquid ratio (S/L = 5, 10, or 15). The extraction was performed irradiating at a potency of 580 W without agitation and for a total extraction time of 10, 20, or 30 min. The results of each extraction were analyzed using response surface methodology. The influence of the three process parameters was investigated on total polyphenols yield, total flavonoids yield, antioxidant activity, and reductive capacity as well. The optimal conditions for the highest CBD content and lowest THC content resulted in 47% ethanol concentration, 10 minutes of extraction time, and an S/L ratio of 5. The model was successfully validated by preparing the Cannabis extract under the calculated conditions.

Alongside the pharmaceutical uses of Cannabis extracts, hemp seeds are widely employed to produce hemp oil. However, the content of Δ^9-THC in the processed hemp seed oils must be under the limits imposed by the jurisdictions of each State. Indeed, although the hemp seeds produce negligible amounts of THC, their outer surface can be contaminated with the enriched in the phytocannabinoids resin secreted by the seeds' bracts. The presence of Δ^9-THC in the final hemp seed products had led to intoxication symptoms in the final consumers. Thus, nowadays the content of THC in hemp products is tightly regulated. Yang et al. investigated the effectiveness of various chemical procedures for the extraction of Δ^9-THC from three brands of hemp seeds and how the extractive methods could influence their commercialization [72]. Four extraction methods were employed, namely, i) microwave extraction, ii) sonication, iii) SE and iv) SFE. As already investigated by Kore et al., the extraction was performed in ethanol at 150°C with stirring, obtaining a complete conversion of CBDA and THCA into the corresponding neutral form. Hemp seeds were macerated in a mortar, transferred into a microwave vessel, and suspended in ethanol. The suspension was irradiated at 150°C with stirring for 20 min in a closed vessel. The yield of the resin (27–38%) achieved was comparable to the other three extraction procedures. In contrast, SE provided higher yields of Δ^9-THC and CBD than the other procedures, resulting in a more robust and appropriate extraction methodology for the testing of hemp seed products. Since the same solvent was used in all the compared extractions, the differences in the number of phytocannabinoids can be attributed to the extraction methods themselves. The results suggest that prolonged heating and solvent cycling in extracting phytocannabinoids from lipid-rich materials such as hemp seeds is mandatory.

5. Conclusion

MASE has rapidly risen during the latest decades as a method for the extraction of secondary metabolites or compounds of pharmaceutical and nutraceutical interest. The use of microwave can generate peculiar, and otherwise impossible to reach extraction mechanisms. As a result, a reduction of the extraction time, improvement of the extraction efficiency, high reproducibility, and robustness of the procedure can be achieved. An increase of the sample throughput is in addition possible, thus it can be considered as the elective technique when a high number of samples have to be processed specially during the first stage of the NADD process, and for evaluating the quality of the natural matrices [9, 29]. For these reasons, MASE has proven to be effective in all aspects, including economical and practical, compared to traditional extraction techniques, especially over SE. Conversely, in MASE the development of the method must be carefully assessed, and all the variables and factors described above must be thoroughly considered to provide some extraction selectivity. Hence,

DoE, response surface methodology, and other statistical approaches are of great help to quickly determine the best conditions to achieve the highest yield of the metabolite of interest from the natural source. However, in the past year, the application of MASE in scalable industrial processes has always encountered several limitation due to the presence of some technological barriers, mainly related to the design of safe instrumentation. Thanks to the technological progress witnessed in recent years, the first industrial-scale ovens finally became commercially available [73, 74].

Food, pharmaceutical, and nutraceutical industries would be benefited from this emerging technology of MASE, which is an excellent substitute for traditional methods such as SE, and other environmentally benign technologies. The promise to be the technique that can respond to the necessities in this field will make MASE the extraction method of choice for the next years.

Conflict of interest

The authors declare no conflict of interest.

Author details

Valeria Cavalloro[1], Emanuela Martino[1], Pasquale Linciano[2] and Simona Collina[2]*

1 Department of Earth and Environmental Science, University of Pavia, Pavia, Italy

2 Department of Drug Science, University of Pavia, Pavia, Italy

*Address all correspondence to: simona.collina@unipv.it

References

[1] Medina-Franco JL. Discovery and Development of Lead Compounds from Natural Sources Using Computational Approaches. In: Evidence-Based Validation of Herbal Medicine. Elsevier; 2015. p. 455-475.

[2] Newman DJ, Cragg GM. Natural Products as Sources of New Drugs over the Nearly Four Decades from 01/1981 to 09/2019. J Nat Prod. 2020 Mar;83(3):770-803.

[3] Herbal Medicine Market Value to Surpass USD 129 Billion Revenue Mark by 2023 at 5.88% CAGR, Predicts Market Research Future. Market Research Future GlobeNewswire. 2019.

[4] Hert J, Irwin JJ, Laggner C, Keiser MJ, Shoichet BK. Quantifying Biogenic Bias in Screening Libraries. Nat Chem Biol. 2009;5(7):479-83.

[5] Popovici V, Bucur LA, Schröder V, Gherghel D, Mihai CT, Caraiane A, et al. Evaluation of the Cytotoxic Activity of the Usnea barbata (L.) F. H. Wigg Dry Extract. Molecules. 2020 Apr;25(8):1865.

[6] Freysdottir J, Omarsdottir S, Ingólfsdóttir K, Vikingsson A, Olafsdottir ES. In vitro and in vivo immunomodulating effects of traditionally prepared extract and purified compounds from *Cetraria islandica*. Int Immunopharmacol. 2008;

[7] McCauley EP, Piña IC, Thompson AD, Bashir K, Weinberg M, Kurz SL, et al. Highlights of marine natural products having parallel scaffolds found from marine-derived bacteria, sponges, and tunicates. J Antibiot (Tokyo). 2020 Aug;73(8):504-25.

[8] Verpoorte R, van der Heijden R, Memelink J. Engineering the plant cell factory for secondary metabolite production. Transgenic Res. 2000;9(4-5):323-43; discussion 321.

[9] Martino E, Della Volpe S, Cavalloro V, Amri B, Kaab LBB, Marrubini G, et al. The use of a microwave-assisted solvent extraction coupled with HPLC-UV/PAD to assess the quality of *Marrubium vulgare* L. (white horehound) herbal raw material. Phytochem Anal. 2019;30(4).

[10] Borges A, José H, Homem V, Simões M. Comparison of Techniques and Solvents on the Antimicrobial and Antioxidant Potential of Extracts from *Acacia dealbata* and *Olea europaea*. Antibiotics. 2020 Jan;9(2):48.

[11] Zhang Q-W, Lin L-G, Ye W-C. Techniques for extraction and isolation of natural products: a comprehensive review. Chin Med. 2018 Dec;13(1):20.

[12] Li Y, Fabiano-Tixier AS, Vian MA, Chemat F. Solvent-free microwave extractionof bioactive compounds providesa tool for green analytical chemistry. Trends Anal Chem. 2013;47:1-11.

[13] Vinatoru M, Mason TJ, Calinescu I. Trends in Analytical Chemistry Ultrasonically assisted extraction (UAE) and microwave assisted extraction (MAE) of functional compounds from plant materials. Trends Anal Chem. 2017;97:159-78.

[14] Xue H, Xu H, Wang X, Shen L, Liu H, Liu C, et al. Effects of Microwave Power on Extraction Kinetic of Anthocyanin from Blueberry Powder considering Absorption of Microwave Energy. 2018;2018.

[15] Vats T, Mishra A. Sustainable Syntheses with Microwave Irradiation. Encycl Inorg Bioinorg Chem. 2016;

[16] Marco B, Agnese C, Giuseppe T. Quality Preservation and Cost

Effectiveness in the Extraction of Nutraceutically-Relevant Fractions from Microbial and Vegetal Matrices. In: Scientific, Health and Social Aspects of the Food Industry. InTech; 2012.

[17] Chan C-H, Yusoff R, Ngoh G-C, Kung FW-L. Microwave-assisted extractions of active ingredients from plants. J Chromatogr A. 2011 Sep;1218(37):6213-25.

[18] Ibrahim NA, Zaini MAA. Solvent selection in microwave assisted extraction of castor oil. Chem Eng Trans. 2017;56:865-70.

[19] Xu W, Chu K, Li H, Zhang Y, Zheng H, Chen R, et al. Ionic Liquid-Based Microwave-Assisted Extraction of Flavonoids from Bauhinia championii (Benth.) Benth. Molecules. 2012 Dec;17(12):14323-35.

[20] Du F-Y, Xiao X-H, Xu P-P, Li G-K. Ionic liquid-based microwave-assisted extraction and HPLC analysis of dehydrocavidine in corydalis saxicola Bunting. Acta Chromatogr [Internet]. 2010 Sep;22(3):459-71. Available from: https://akjournals.com/doi/10.1556/achrom.22.2010.3.9

[21] Vigani B, Rossi S, Gentile M, Sandri G, Bonferoni MC, Cavalloro V, et al. Development of a mucoadhesive and an in situ gelling formulation based on κ-carrageenan for application on oral mucosa and esophagus walls. II. Loading of a bioactive hydroalcoholic extract. Mar Drugs. 2019;17(3).

[22] Martino E, Ramaiola I, Urbano M, Bracco F, Collina S. Microwave-assisted extraction of coumarin and related compounds from *Melilotus officinalis* (L.) Pallas as an alternative to Soxhlet and ultrasound-assisted extraction. J Chromatogr A. 2006 Sep;1125(2):147-51.

[23] Xue H, Xu H, Wang X, Shen L, Liu H, Liu C, et al. Effects of Microwave Power on Extraction Kinetic of Anthocyanin from Blueberry Powder considering Absorption of Microwave Energy. J Food Qual. 2018;2018:1-13.

[24] Routray W, Orsat V. Microwave-Assisted Extraction of Flavonoids: A Review. Food Bioprocess Technol. 2012 Feb;5(2):409-24.

[25] Pimentel-Moral S, Borrás-Linares I, Lozano-Sánchez J, Arráez-Román D, Martínez-Férez A, Segura-Carretero A. Microwave-assisted extraction for *Hibiscus sabdariffa* bioactive compounds. J Pharm Biomed Anal. 2018 Jul;156:313-22.

[26] N. Politis S, Colombo P, Colombo G, M. Rekkas D. Design of experiments (DoE) in pharmaceutical development. Drug Dev Ind Pharm [Internet]. 2017 Jun 3;43(6):889-901. Available from: https://www.tandfonline.com/doi/full/10.1080/03639045.2017.1291672

[27] Camel V. Microwave-assisted solvent extraction of environmental samples. TrAC Trends Anal Chem [Internet]. 2000 Apr;19(4):229-48. Available from: https://linkinghub.elsevier.com/retrieve/pii/S0165993699001855

[28] Sanchez-Prado L, Garcia-Jares C, Dagnac T, Llompart M. Microwave-assisted extraction of emerging pollutants in environmental and biological samples before chromatographic determination. TrAC Trends Anal Chem [Internet]. 2015 Sep;71:119-43. Available from: https://linkinghub.elsevier.com/retrieve/pii/S0165993615001442

[29] Granata MU, Bracco F, Catoni R, Cavalloro V, Martino E. Secondary metabolites profile and physiological leaf traits in wild and cultivated *Corylus avellana* under different nutritional status. Nat Prod Res. 2019 Oct;1-8.

[30] Martino E, Collina S, Rossi D, Bazzoni D, Gaggeri R, Bracco F, et al.

Influence of the extraction mode on the yield of hyperoside, vitexin and vitexin-2″- O -rhamnoside from *Crataegus monogyna* Jacq. (hawthorn). Phytochem Anal. 2008 Nov;19(6):534-40.

[31] He M, Min J-W, Kong W-L, He X-H, Li J-X, Peng B-W. A review on the pharmacological effects of vitexin and isovitexin. Fitoterapia [Internet]. 2016 Dec;115:74-85. Available from: https://linkinghub.elsevier.com/retrieve/pii/S0367326X16304488

[32] Tang X, Zhu D, Huai W, Zhang W, Fu C, Xie X, et al. Simultaneous extraction and separation of flavonoids and alkaloids from Crotalaria sessiliflora L. by microwave-assisted cloud-point extraction. Sep Purif Technol [Internet]. 2017 Mar;175:266-73. Available from: https://linkinghub.elsevier.com/retrieve/pii/S1383586616307110

[33] Madej K. Microwave-assisted and cloud-point extraction in determination of drugs and other bioactive compounds. TrAC Trends Anal Chem. 2009 Apr;28(4):436-46.

[34] Belwal T, Pandey A, Bhatt ID, Rawal RS. Optimized microwave assisted extraction (MAE) of alkaloids and polyphenols from Berberis roots using multiple-component analysis. Sci Rep [Internet]. 2020 Dec 22;10(1):917. Available from: http://www.nature.com/articles/s41598-020-57585-8

[35] Pellati F, Benvenuti S. Determination of ephedrine alkaloids in Ephedra natural products using HPLC on a pentafluorophenylpropyl stationary phase. J Pharm Biomed Anal. 2008 Sep;48(2):254-63.

[36] Brachet A, Christen P, Veuthey J-L. Focused microwave-assisted extraction of cocaine and benzoylecgonine from coca leaves. Phytochem Anal. 2002 May;13(3):162-9.

[37] Nowak J, Woźniakiewicz M, Klepacki P, Sowa A, Kościelniak P. Identification and determination of ergot alkaloids in Morning Glory cultivars. Anal Bioanal Chem. 2016 May;408(12):3093-102.

[38] Petruczynik A. Analysis of alkaloids from different chemical groups by different liquid chromatography methods. Open Chem [Internet]. 2012 Jun 1;10(3):802-35. Available from: https://www.degruyter.com/view/journals/chem/10/3/article-p802.xml

[39] Xie D-T, Wang Y-Q, Kang Y, Hu Q-F, Su N-Y, Huang J-M, et al. Microwave-assisted extraction of bioactive alkaloids from Stephania sinica. Sep Purif Technol. 2014 Jun;130:173-81.

[40] Ganesapillai M, Singh A, Subba Rao B, De D, Juneja U. EXTRACTION OF ALKALOIDS FROM MICROWAVE DRIED ADATHODA VASICA LEAVES - A COMPARATIVE STUDY. Int J Pharma Bio Sci. 2015;6(1):121-39.

[41] Pan X, Niu G, Liu H. Microwave-assisted extraction of tea polyphenols and tea caffeine from green tea leaves. Chem Eng Process Process Intensif [Internet]. 2003 Feb;42(2):129-33. Available from: https://linkinghub.elsevier.com/retrieve/pii/S0255270102000375

[42] Xiong W, Chen X, Lv G, Hu D, Zhao J, Li S. Optimization of microwave-assisted extraction of bioactive alkaloids from lotus plumule using response surface methodology. J Pharm Anal. 2016 Dec;6(6):382-8.

[43] Zhou S, Wu X, Huang Y, Xie X, Lin Y, Fan H, et al. Microwave-assisted aqueous two-phase extraction of alkaloids from Radix Sophorae Tonkinensis with an ethanol/Na2HPO4 system: Process optimization, composition identification and quantification analysis. Ind Crops Prod. 2018 Oct;122:316-28.

[44] Fernández P, Lago M, Lorenzo RA, Carro AM, Bermejo AM, Tabernero MJ. Microwave assisted extraction of drugs of abuse from human urine. J Appl Toxicol [Internet]. 2007 Jul;27(4):373-9. Available from: http://doi.wiley.com/10.1002/jat.1216

[45] Fernández P, Lago M, Lorenzo RA, Carro AM, Bermejo AM, Tabernero MJ. Optimization of a rapid microwave-assisted extraction method for the simultaneous determination of opiates, cocaine and their metabolites in human hair. J Chromatogr B [Internet]. 2009 Jun;877(18-19):1743-50. Available from: https://linkinghub.elsevier.com/retrieve/pii/S1570023209003055

[46] Fernández P, Seoane S, Vázquez C, Bermejo AM, Carro AM, Lorenzo RA. A rapid analytical method based on microwave-assisted extraction for the determination of drugs of abuse in vitreous humor. Anal Bioanal Chem [Internet]. 2011 Oct 7;401(7):2177-86. Available from: http://link.springer.com/10.1007/s00216-011-5279-6

[47] Rimando AM, Kalt W, Magee JB, Dewey J, Ballington JR. Resveratrol, Pterostilbene, and Piceatannol in Vaccinium Berries. J Agric Food Chem. 2004 Jul;52(15):4713-9.

[48] Kim H-K, Do J-R, Lim T-S, Akram K, Yoon S-R, Kwon J-H. Optimisation of microwave-assisted extraction for functional properties of Vitis coignetiae extract by response surface methodology. J Sci Food Agric. 2012 Jun;92(8):1780-5.

[49] Piñeiro Z, Marrufo-Curtido A, Vela C, Palma M. Microwave-assisted extraction of stilbenes from woody vine material. Food Bioprod Process. 2017 May;103:18-26.

[50] Hao J, Han W, Huang S, Xue B, Deng X. Microwave-assisted extraction of artemisinin from *Artemisia annua* L. Sep Purif Technol. 2002 Sep;28(3):191-6.

[51] Giampietro L, Angelo AD, Giancristofaro A, Ammazzalorso A, Filippis B De, Matteo M Di, et al. Effect of Stilbene and Chalcone Scaffolds Incorporation in Clofibric Acid on PPAR Agonistic Activity. Med Chem (Los Angeles). 2014;10:59-65.

[52] De Filippis B, Linciano P, Ammazzalorso A, Di Giovanni C, Fantacuzzi M, Giampietro L, et al. Structural development studies of PPARs ligands based on tyrosine scaffold. Eur J Med Chem. 2015;89:817-25.

[53] Mei Y-Z, Liu R-X, Wang D-P, Wang X, Dai C-C. Biocatalysis and biotransformation of resveratrol in microorganisms. Biotechnol Lett. 2015 Jan;37(1):9-18.

[54] Quideau S, Deffieux D, Douat-Casassus C, Pouységu L. Plant Polyphenols: Chemical Properties, Biological Activities, and Synthesis. Angew Chemie Int Ed. 2011 Jan;50(3):586-621.

[55] Saraswati S. Velu, Noel F. Thomas, Jean-Frederic F. Weber. Strategies and Methods for the Syntheses of Natural Oligomeric Stilbenoids and Analogues. Curr Org Chem. 2012 Apr;16(5):605-62.

[56] Donnez D, Jeandet P, Clément C, Courot E. Bioproduction of resveratrol and stilbene derivatives by plant cells and microorganisms. Trends Biotechnol. 2009 Dec;27(12):706-13.

[57] Nopo-Olazabal C, Hubstenberger J, Nopo-Olazabal L, Medina-Bolivar F. Antioxidant Activity of Selected Stilbenoids and Their Bioproduction in Hairy Root Cultures of Muscadine Grape (*Vitis rotundifolia* Michx.). J Agric Food Chem. 2013 Dec;61(48):11744-58.

[58] Almagro L, Belchí-Navarro S, Sabater-Jara AB, Vera-Urbina JC, Sellés-Marchart S, Bru R, et al. Bioproduction of trans-Resveratrol from Grapevine Cell Cultures. In: Natural Products. Berlin, Heidelberg: Springer Berlin Heidelberg; 2013. p. 1683-713.

[59] Garcia L, Garcia R, Pacheco G, Sutili F, Souza R De, Mansur E, et al. Optimized Extraction of Resveratrol from Arachis repens Handro by Ultrasound and Microwave: A Correlation Study with the Antioxidant Properties and Phenol Contents. Sci World J. 2016;2016:1-10.

[60] Dang Y-Y, Zhang H, Xiu Z-L. Microwave-assisted aqueous two-phase extraction of phenolics from grape (*Vitis vinifera*) seed. J Chem Technol Biotechnol. 2014 Oct;89(10):1576-81.

[61] Kim J-S, Ha T-Y, Ahn J, Kim H-K, Kim S. Pterostilbene from Vitis coignetiae protect H2O2-induced inhibition of gap junctional intercellular communication in rat liver cell line. Food Chem Toxicol. 2009 Feb;47(2):404-9.

[62] Liu C-Z, Zhou H-Y, Zhao Y. An effective method for fast determination of artemisinin in *Artemisia annua* L. by high performance liquid chromatography with evaporative light scattering detection. Anal Chim Acta. 2007 Jan;581(2):298-302.

[63] Mattina MJI, Berger WAI, Denson CL. Microwave-Assisted Extraction of Taxanes from Taxus Biomass. J Agric Food Chem. 1997 Dec;45(12):4691-6.

[64] Lee S-H, Kim J-H. Kinetic and thermodynamic characteristics of microwave-assisted extraction for the recovery of paclitaxel from Taxus chinensis. Process Biochem. 2019 Jan;76:187-93.

[65] Citti C, Linciano P, Forni F, Vandelli MA, Gigli G, Laganà A, et al. Analysis of impurities of cannabidiol from hemp. Isolation, characterization and synthesis of cannabidibutol, the novel cannabidiol butyl analog. J Pharm Biomed Anal. 2019;175.

[66] Linciano P, Citti C, Luongo L, Belardo C, Maione S, Vandelli MA, et al. Isolation of a High-Affinity Cannabinoid for the Human CB1 Receptor from a Medicinal *Cannabis sativa* Variety: Δ9-Tetrahydrocannabutol, the Butyl Homologue of Δ9-Tetrahydrocannabinol. J Nat Prod. 2020;83(1):88-98.

[67] Citti C, Linciano P, Russo F, Luongo L, Iannotta M, Maione S, et al. A novel phytocannabinoid isolated from *Cannabis sativa* L. with an in vivo cannabimimetic activity higher than Δ9-tetrahydrocannabinol: Δ9-Tetrahydrocannabiphorol. Sci Rep. 2019;9(1):1-13.

[68] Nahar L, Uddin SJ, Alam MA, Sarker SD. Extraction of naturally occurring cannabinoids: an update. Phytochem Anal. 2020 Sep;pca.2987.

[69] Lewis-Bakker MM, Yang Y, Vyawahare R, Kotra LP. Extractions of Medical Cannabis Cultivars and the Role of Decarboxylation in Optimal Receptor Responses. Cannabis Cannabinoid Res. 2019 Sep;4(3):183-94.

[70] Kotra LP, Lewis MM, Wasilewski E, Grover H. Decarboxylated Cannabis Resins, Uses Thereof and Methods of Making Same. WO/2018/000094, 2018.

[71] Drinić Z, Vladić J, Koren A, Zeremski T, Stojanov N, Kiprovski B, et al. Microwave-assisted extraction of cannabinoids and antioxidants from *Cannabis sativa* aerial parts and process modeling. J Chem Technol Biotechnol. 2020 Mar;95(3):831-9.

[72] Yang Y, Lewis MM, Bello AM, Wasilewski E, Clarke HA, Kotra LP. Cannabis sativa (Hemp) Seeds, Δ 9 -Tetrahydrocannabinol, and Potential Overdose. Cannabis Cannabinoid Res. 2017 Jan;2(1):274-81.

[73] Li Y, Radoiu M, Fabiano-Tixier A-S, Chemat F. From Laboratory to Industry: Scale-Up, Quality, and Safety Consideration for Microwave-Assisted Extraction. In 2012. p. 207-29.

[74] Périno S, Pierson JT, Ruiz K, Cravotto G, Chemat F. Laboratory to pilot scale: Microwave extraction for polyphenols lettuce. Food Chem. 2016;204:108-14.

Chapter 3

Microwave Synthesized Functional Dyes

Sheetal Marganakop, Pramod Kattimani, Sudha Belgur Satyanarayana and Ravindra Kamble

Abstract

Microwave chemistry involves the application of microwave radiation to chemical reactions and has played an important role in organic synthesis. Functional dyes are those with hi-tech applications and this chapter attempts to provide an overview of the recent developments in microwave-assisted synthesis of functional dyes. Emphasis has been paid to the microwave-assisted synthesis of dye molecules which are useful in hi-tech applications such as optoelectronics (dye-sensitized solar cells), photochromic materials, liquid crystal displays, newer emissive displays (organic-light emitting devices), electronic materials (organic semiconductors), imaging technologies (electrophotography viz., photocopying and laser printing), biomedical applications (fluorescent sensors and anticancer treatment such as photodynamic therapy). In this chapter, the advantages of microwaves as a source of energy for heating synthesis reactions have been demonstrated. The use of microwaves to functional dyes is a paradigm shift in dye chemistry. Until recently most academic laboratories did not practice this technique in the synthesis of such functional dyes but many reports are being appeared in the journals of high repute.

Keywords: microwave-assisted organic synthesis, functional dyes, solar cells, fluorescent sensors, organic-light emitting diodes, photochromic materials

1. Introduction

Microwaves are the portion of the electromagnetic spectrum with the wavelengths from 1 mm to 1 m with corresponding frequencies between 300 MHz and 300 GHz. The frequencies used for cellular phones, radar, and television satellite communications are within this portion of the electromagnetic spectrum [1]. Microwaves have been employed in a non-classical heating technique which is popularly known as "Bunsen Burner of the 21st century" and has attained enormous importance since many materials (solids or liquids) can transform electromagnetic energy into heat. The microwave-assisted organic synthesis (MAOS) has made revolutionary changes in the methodology since there is a dramatic enhancement in the yield of the reaction, modifications of selectivity, increased purity of products, simplified work-up procedure, and above all reduction in the reaction time. These are the primary benefits over conventional methods. The microwave technique has been applied efficiently in the organic synthesis, polymer chemistry, material sciences, nanotechnology, biochemical processes, thermal food processing, hydrothermal and solvothermal processing, etc. [2]. The energy efficiency is higher in the case of microwave heating in comparison with the conventional heating as evidenced by one such Suzuki reaction in which there is an 85 fold reduction in energy demand when compared to a reaction on an oil bath and a microwave reactor [3].

During a chemical reaction under the conventional heating, the energy is introduced by convection, conduction, and radiation of heat from the surfaces of the reactants in the solution, and the energy transfer occurs due to thermal gradients. But in the case of the microwave irradiation, the energy is introduced through the electromagnetic field interaction into the molecules and the transfer of electromagnetic energy to thermal energy is energy conversion instead of heat transfer. This variation in the mode of introduction of energy leads to the advantages of using microwaves during chemical reactions. The microwaves penetrate easily into the bulk and, hence, heat evolves throughout the volume of the reaction mixture. As a result, fast and uniform heating of the reaction mixture can be advanced. In conventional heating, it is necessary to slow rates of heating to minimize the steep thermal gradients and obviate the process-induced stresses. As microwaves can transfer energy into all volumes of the reaction mixture, the potential exists to reduce the processing time and enhance the overall quality [4].

Although the use of microwaves for organic synthesis is widespread, the documentation of this technology to the synthesis of the functional dyes is a relatively new development. The use of microwave energy for their synthesis has the potential to offer similar advantages in reduced reaction times and energy savings for obtaining useful materials such as dyes possessing hi-tech applications.

2. Functional dyes

Color plays an important role in the world in which we are living. Color can sway thinking, change actions, and cause reactions. If properly used, color can even save on energy consumption. The colors are characterized by their ability to absorb light in the visible spectrum (from 380 to 750 nm). The dyeing industry is in existence since 2000 years BCE wherein dyes were obtained from natural sources *viz.*, plants, insects/animals, and mineral [5, 6]. A drastic development occurred after the discovery of the dye Mauveine by W.H. Perkin in 1856 while trying to synthesize quinine [7]. Dyes are the organic compounds with three essential groups in their molecules *viz.*, the chromophore, the auxochrome, and the matrix. The chromophore is an active site of the dye which may be an atom or group whose presence is responsible for the color of a dye. The auxochrome is responsible for the intensity of the color of the dye with lone pairs of electrons.

It was Yoshida and Kato who used the term "functional dye" for the first time in 1981 due to the advancements and growth of dye chemistry related to high-technology (hi-tech) applications that are divergent from the well known traditional applications [8]. Hi-tech applications of dyes include the fields *viz.*, optoelectronics (i.e. Dye-sensitized solar cells), photochemical materials, liquid crystal displays (LCD), and the newer emissive displays i.e. organic light-emitting diodes (O-LED), electronic materials (organic semiconductors), imaging technologies (electrophotography which includes photocopying and laser printing), thermal printing, and especially ink-jet printing, biotechnology (in dye-affinity chromatography for the purification of proteins and enzymes), biomedical applications (fluorescent sensors and anticancer treatments such as photodynamic therapy). All these fields were responsible for the design and synthesis of newer dyes to meet new and demanding criteria. Dyes, and related ultraviolet and particularly infrared active molecules, which have been specifically designed for these hi-tech applications, are called functional dyes.

Common dyes have been synthesized by applying mainly the conventional methods and also by microwave assistance. In the following sections the functional dyes used in solar cells, fluorescent sensors, fluorescent dyes to print on fibres, photochromic materials, O-LEDs, and dyes with advanced applications which were synthesized only under microwave irradiation are discussed.

2.1 Dyes (sensitizers) used in solar cells

2.1.1 Dye-sensitized solar cells (DSSCs)

To prevent harmful impact on the environment by conventional energy sources it is necessary to use the alternative energy sources, specially, the solar cells. The conversion of sunlight into electricity is a clean, abundant, and renewable energy source. The amount of energy available from the sun to the earth is of the order of 3×10^{24} joules/year thus making it the best among sustainable energies. Photovoltaic devices have been fabricated using inorganic materials of high purity and energy-intensive processing techniques. The fabrication using these inorganic materials is not economical and often used scarce toxic materials. Therefore, such solid-state junction devices have been challenged by the 3rd generation dye-sensitized solar cells (DSSCs) which are based on interpenetrating network structures containing metal-free organic dyes as sensitizers [9].

In the conventional systems, the semiconductor does the task of light absorption as well as charge carrier transport. However, these two functions are separated in DSSCs by the metal-free organic dye and TiO_2 in presence of an electrolyte. Hence, new ways of manufacturing the solar cells that can be scaled economically up to large volumes are essential. In this regard, a new generation of DSSCs also known as "Grätzel cells" has been fabricated by O'Regan and Grätzel [10]. A Grätzel cell consists of nanoporous titanium dioxide applied on transparent conducting oxide which is further made to absorb the dye from its solution. This film loaded with dye/sensitizer is immersed in an electrolyte containing a redox couple and placed on a platinum counter electrode. After irradiation, the excited electron from the dye (sensitizer) is transferred to the conduction band of TiO_2 and diffuses through its porous network to the contact. Thus oxidized dye is further reduced to the original state by the supply of electrons through a liquid electrolyte redox couple within the pores [11].

The organic dye sensitizers consist of three important components *viz.*, electron donor (D), π-conjugated spacer (π), and electron acceptor (A). Electron acceptors are generally acid ligands which also act as anchoring groups for loading the dye on TiO_2 surface. The π-conjugated spacer (*viz.*, conjugated double bonds, phenyl rings, thiophene, polythiophenes, etc) acts as a bridge to transfer electrons between the donor and the acceptor group and it is the key part which can induce a shift of both the highest occupied molecular orbital (HOMO) and the lowest unoccupied molecular orbital (LUMO) levels so that the photophysical properties may be tuned. The organic dyes/sensitizers belong to different classes depending on the donors such as triphenylamine, phenothiazine, fluorene, coumarin, carbazoles, etc. which have been profusely synthesized, and their power conversion efficiency as sensitizers have been reported and reviewed exclusively [12]. The structures of the dyes/ sensitizers synthesized under microwave irradiation along with the parameters such as short-circuit current (*Jsc*), open-circuit voltage (*Voc*), Fill Factor (FF), and power conversion efficiency (PCE) of the solar cells fabricated using these dyes are discussed.

2.1.2 Microwave synthesized dyes/sensitizers in DSSCs

Novel donor-π-acceptor (D-π–A) dyes bearing the pyrimidine unit as an electron acceptor appended to thiophene and carbazole unit **1a-c** were obtained by a combination of two processes *viz*., the microwave-assisted Suzuki cross-coupling reaction and nucleophilic aromatic substitution of hydrogen (S_NH) [13]. Among these dyes, **1b** was used as a photosensitizer in a fabricated solar cell since this dye showed a maximum extinction coefficient. The short-circuit current density (*Jsc*) was 2.04 mA cm^{-2}, and the open-circuit voltage (*Voc*) observed was 0.525 V. The calculated power conversion efficiency (PCE) of the cell (η) was 0.91 at a fill factor (FF) of 0.85. A series of dithienosilole-based terpolymers **2a-e** as sensitizers have been synthesized. Different dithienosilole monomers were used along with nonanoyl group and malononitrile as the electron acceptor *via* microwave-assisted Stille coupling polymerization to obtain the polymer sensitizers **2a-e**.

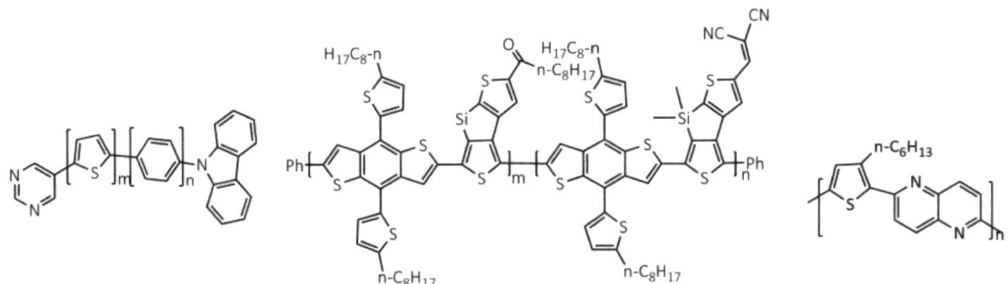

1a: m = 1, n = 1; **1b**: m = 1, n = 2; **1c**: m = 2, n = 1

2a; m:n = 0:1, **2b**; m:n = 1:3, **2c**; m:n = 1:1, **2d**; m:n = 3:1, **2e**; m:n = 1:0

3

The devices obtained using these sensitizers **2a-e** exhibited the high open-circuit voltage (*Voc*) of 1.00–1.06 V which can be attributed to the low-lying HOMO energy levels. Sensitizer **2e** showed the best PCE of 2.32% (*Voc* = 1.06 V, *Jsc* = 5.92 mA/cm^2 and FF = 0.39) which is due to the components of the conjugated backbones and play a pivotal role in their photovoltaic performance. When the polymerisation process was optimized i.e. in polymer **2e** with higher molecular weight (Mn = 23.3 kDa) an increased PCE of 3.29% (*Voc* = 1.07 V, *Jsc* = 7.53 mA/cm^2 and FF = 0.41) was observed [14]. The above reports showed the outstanding thermal stabilities and electrical conductivity of polythiophenes. Hence, a semiconducting polymer *viz*., poly[1,5-naphthyridine-(3-hexylthiophene)] **3** was prepared by microwave-assisted *Suzuki-Miyura* cross-coupling reaction using 3-hexylthiophene-2,5-diboronic ester and 2,6-dibromo-1,5-naphthyridine [15]. This polymer **3** was used as a photosensitizer in a fabricated solar cell. The solar cell so prepared was illuminated under AM 1.5 G at 100 mW/cm^2 which showed a PCE of 0.67% with an open-circuit voltage of (Voc) 621 mV, a short-circuit current of 2.0 mA/cm^2, and a FF of 55%.

Three push-pull Donor-π-Acceptor structured dyes **4**, **5** and **6** having imidazo [1,2-*a*]pyridine heterocycles as additional π-conjugated linker was synthesized. Triphenylamine (TPA) was introduced as an electron-donor unit and cyanoacetic acid through thiophene as linker **4** and **5** or double rhodanine acetic acid **6** were employed as anchoring groups in different positions of the heterocyclic core [16]. DSSC devices with these dyes **4**–**6** were assembled and tested using different electrolytes and dye baths. The best efficiencies were obtained for dye **4** i.e. *Jsc* 2.34 (mA/cm^2), *Voc* 650 mV, FF 0.42, η (%) 0.64 and for **5** *Jsc* 2.14 (mA/cm^2), Voc 502 mV, FF 0.42, η (%) 0.45.

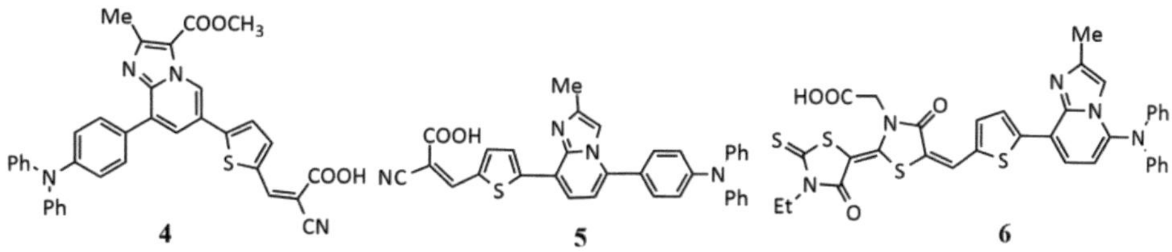

4 **5** **6**

Due to inefficient electron injection from HOMO to TiO_2 conduction band or dye aggregation leading to a potential barrier the dye **6** showed the lowest efficiency irrespective of dye bath solvent and electrolytes [17].

In view of the importance of thiophene as the significant moiety in the design of polymer-based sensitizers, narrow band gap conjugated polymer **7** was obtained from 4,6-bis(4-tetradecylthien-2-yl)thieno[3,4-*c*]thiadiazole, and thieno[3,2-*b*] thiophene using Stille coupling reaction under microwave irradiation. This polymer exhibited good solution processability and absorbed the UV/Vis light from 300 nm to 1260 nm with an optical band gap of 0.98 eV in solid state. Photovoltaic devices using the blend films **8**, **9** from **7** and [6,6]-phenyl-C_{61} butyric acid methyl ester ($PC_{61}BM$) or [6,6]-phenyl-C_{71} butyric acid methyl ester ($PC_{71}BM$) having the configuration ITO/PEDOT:PSS/blend film/Ca/Al, provided power conversion efficiencies (PCEs) of 0.65%, and 1.12% respectively with light response from 300 nm to 1260 nm under AM 1.5 G with irradiation of 100 mW cm^{-2} [18].

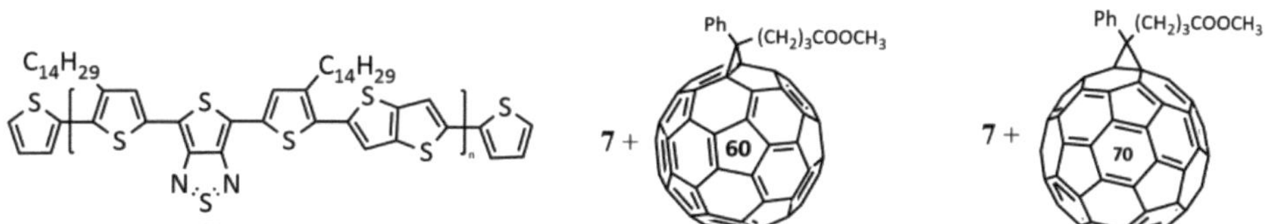

7 **8** Polymer blends **9**

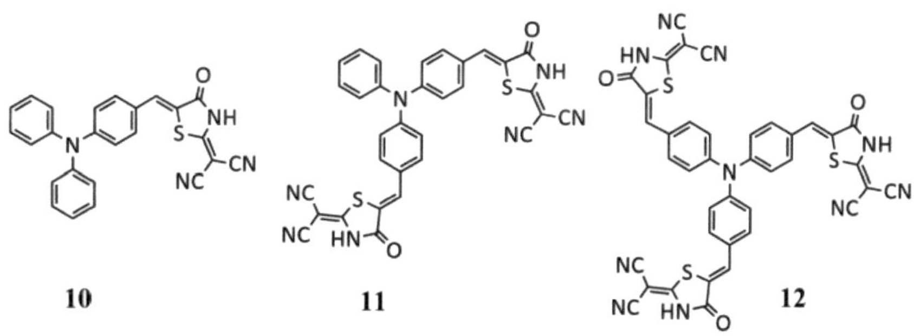

10 **11** **12**

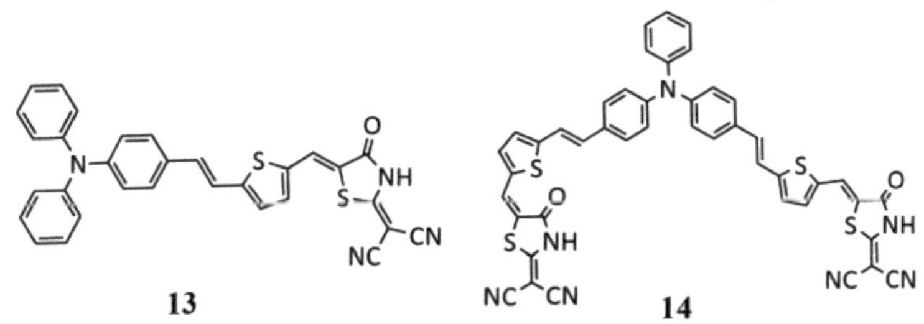

13 **14**

Triphenylamine based dye sensitizers **10–14** were prepared under microwave irradiation by incorporating 2-(1,1-dicyanomethylene) rhodanine which acts both as electron acceptor as well as anchoring group on titanium dioxide [19]. Triphenylamine and vinyl thiophene are the donors and the π spacers. The dye containing two 2-(1,1-dicyanomethylene)rhodanine units and no thiophene units i.e. dye **11** showed the best photovoltaic performance with a short-circuit photocurrent density (*Jsc*) of 7.76 mA/cm^2, an open-circuit photovoltage of 0.62 V, and a fill factor of 0.68, corresponding to an overall conversion efficiency of 3.78% under AM 1.5 irradiation (100 mW/cm^2). The *J*sc of the solar cells fabricated using these dyes increased in the order of **13** < **11** < **10, 12** < **14**.

The general design of organic dye sensitizers is usually in the order D-π-A. However, molecular conjugated chromophores combining only electron donor (D) and acceptor (A) blocks have also been designed and synthesized as active materials for organic solar cells [20]. In view of this, D-A-D dyes **15, 16** and **17** obtained by reaction of mono-formyl triarylamines with 2,3-diaminomaleonitrile. Such D-A-D dyes are expected to show absorption of two photons hence may be used in dyes and solar cells, and focused on their potentialities as a donor material in basic planar heterojunction solar cells. These compounds **15, 16,** and **17** have been evaluated as dyes in solar cells ITO/PEDOT-PSS/dye/C60/Al. The open-circuit voltage (Voc), short-circuit current densities (*Jsc*), fill factor (FF), and power conversion efficiencies (PCEs) of cells of 0.28 cm^2 active area were determined under AM 1.5 simulation solar illumination. Compound **17** did not lead to devices of a quality sufficient for evaluation. Fabricated devices obtained from dyes **15** and **16** respectively gave PCEs of 0.70 and 0.53%.

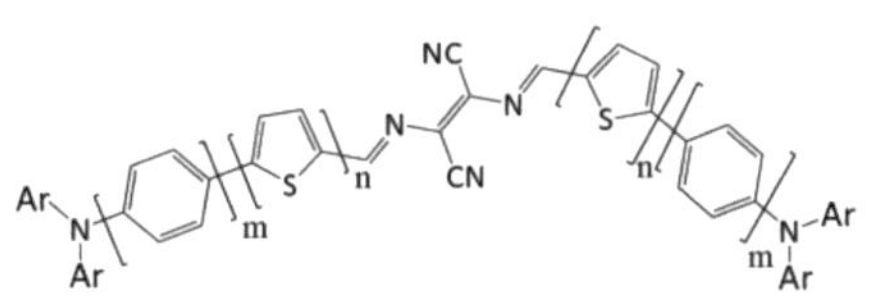

Ar = 4-CH_3O-C_6H_4,

15; m = 1, n = 0, **16**; m= 0, n = 1, **17**; m = 1, n = 1

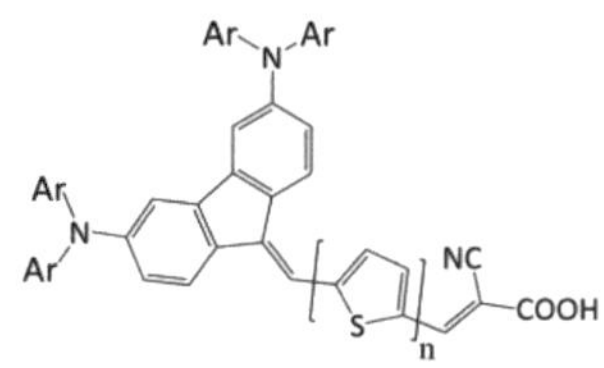

18; Ar = C_6H_5, n = 0, **19**; Ar = C_6H_5, n = 1,
20; R = C_6H_5 n = 2, **21**; R = 4-CH_3O-C_6H_4, n = 1,
22; R = CH_3O-C_6H_4, n = 2, **23**; R = $C_8H_{17}O$-C_6H_4, n = 2

2D-π-A dyes **18–20** comprising of dibenzofulvene-thiophene as π-bridge which is flanked by diarylamine donor groups and cyanoacrylic acid as anchoring as well as acceptor have been synthesized under microwave irradiation [21]. The dye **20** containing two thiophene rings as spacer shows an IPCE action spectrum with a high plateau from 390 nm to 600 nm increased open-circuit photovoltage by 40 mV and short-circuit photocurrent by 7.03 mA cm^{-2}. Using Chenodeoxycholic acid (CDCA) as the co-adsorbent material, the *Jsc* of **22** was increased to 14.98 mA cm^{-2} and a strong enhancement in the overall conversion efficiency (7.45%) was realized by **20** compared to **18** (1.08%) in liquid electrolyte-based DSSCs. This work was further extended by the same research group [22] in which methoxy groups were introduced on the phenyl rings i.e. dyes **21–22** and also long fatty alkyl chain *viz.*, octyloxy was introduced in view of increasing the donor capability and to avoid the aggregation and to increase physical insulation between electrolyte system and the TiO_2 layer *i.e.* dye **23**. These dyes **21–23** exhibit rather similar photophysical properties for the lowest-lying optically active excitations and it was observed that the lowest excitation lay in all cases at 2.12–2.50 eV. Compound **21** showed a promising PCE of 5.90. The structural molecular variations evidenced positive effects on the photovoltaic performances of dyes as proved by PCEs of 7.50% and 7.80% obtained with dyes **22** and **23** respectively.

Donor-acceptor dyes **24–28** based on 3-methylquinoxaline-2(1*H*)one under microwave condition involving Knoevenagel reaction were designed with electron-donor groups such as triphenylamine (TPA) **24–26** ferrocene **25**, N,N-dimethylaminobenzene **27**, and ((*E*)-4,4′-(ethene-1,2-diyl)bis(N,N-diphenyl-aniline)) **28**. The dyes **26** and **27** showed higher power efficiency (0.31 and 0.40 respectively) as expected for their higher values of *Jsc* and *Voc*. This suggests that these structures decrease the recombination processes by preventing the approach of tri-iodide ions to the semiconductor surface, thus decreasing the electron transfer from TiO_2 conduction band to tri-iodide ions electrolyte. The higher efficiency of the dyes **26** and **27** may also be due to the enhanced conjugation of triphenyl-amine units to anchoring amide groups. This has improved the electron injection into semiconductor conduction band which helps in the photovoltaic performance. The remaining dyes did not show significant efficiencies [23].

A new series of oxindole sensitizers (**29–33**) were designed and synthesized under microwave irradiation [24]. These exhibited respectable photoelectric conversion efficiencies due to excellent electron-donating triphenylamine (TPA) donor and the thiophene in the spacer and are differentiated by various halogen-substituted oxindole acceptors. The cell performance was analyzed by fabricating solar cells. The parent dye **29** exhibited *Jsc* = 10.03 $mAcm^{-2}$, *Voc* = 680 mV, and FF = 0.699, corresponding to an overall η = 4.76%. The incorporation of halogen substitutions on the parent dye enhanced the PCEs. Solar cell containing fluoro substituent i.e. dye **30** achieved *Jsc* = 11.32 mA cm^{-2}, *Voc* = 690 mV, and FF = 0.695, corresponding to an overall η = 5.43%, which was approximately 14% higher than that of the non-substituted oxindole sensitizer **29**. The efficiency was increased due to increased *Jsc* which may be attributed to the electronic coupling of *fluoro* substituent in the compound **29** with the anchoring group COOH [25].

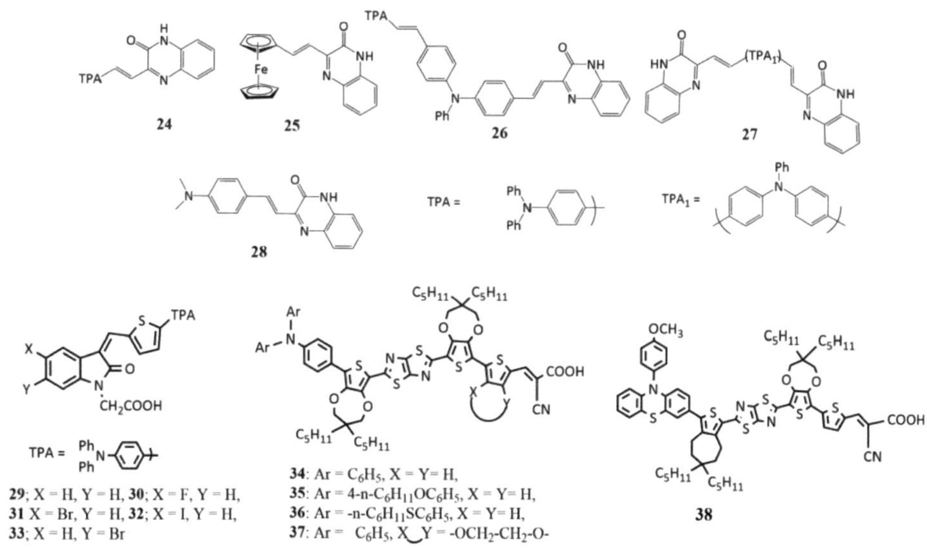

The substitution of fluoro substituent with other halo substituents showed further enhancement in the DSSC performance. Among all the halogen substituted sensitizers, the bromo substituted sensitizer **31** exhibited the highest photovoltaic parameters (*Jsc* = 12.46 mA cm^{-2}, Voc = 720 mV, and FF = 0.708) with an overall conversion efficiency (η) of 6.35%. The improved photocurrent of the sensitizer **31** suggested that compared to fluoro substitution, the bromo substituted dye exhibited better cell performance. Interestingly, the altered position of the substituent with respect to the anchoring group exhibited a negative effect on the solar cell performance. The dye **33** anchored DSSC showed lower current density (*Jsc* = 9.66 mA cm^{-2}, Voc = 630 mV, and FF = 0.690), that is corresponding to an overall η = 4.21% which is due to the absence of electronic coupling of substitution with the anchoring group.

Computationally designed thiazolo [5,4-*d*]thiazole-based D-π-A organic dyes **34–38** have been synthesized [26]. These have been further derivatized with bispentylpropylenedioxythiophene (ProDOT) moieties in the π-spacer and triarylamine and phenothiazine **34–37** and **38** respectively as donors. Bulky and electron-rich ProDOT groups enhanced the physical–chemical properties, including visible light absorption instead of the presence of the electron-poor thiazolothiazole. Small-scale (0.25 cm^2) devices using these dyes **34–38** showed the PCEs up to 7.71%, surpassing those obtained with two different reference dyes. Transparent larger area cells (3.6 cm^2) also showed good η values up to 6.35%, not requiring the use of a co-adsorbent, and retained their initial efficiency over a period of 1000 h storage at 85°C. Following the promising results obtained with small-scale DSSCs (0.25 cm^2), the authors fabricated larger area (3.60 cm^2) strip cells to analyze the effect of increased active surface area on the efficiency and stability. Small-scale solar cells built with **34–38**, both transparent and opaque, gave good power conversion efficiencies (η up to 7.71%), which in the case of dyes **36** and **38** were clearly superior to those obtained with standard Ru-dye Z907. Larger-scale strip cells featuring thin films of transparent TiO_2 (3–5 mm) and a high stability electrolyte, gave efficiencies in line with those obtained with the smaller devices, with dye **36** being once again the best sensitizer (η up to 6.35%).

Two isoindigo-based conjugated polymers **39–40** composed of isoindigo with 2-decyltetradecane (DT) and bithiophene with/without fluorination were prepared under microwave irradiation [27].

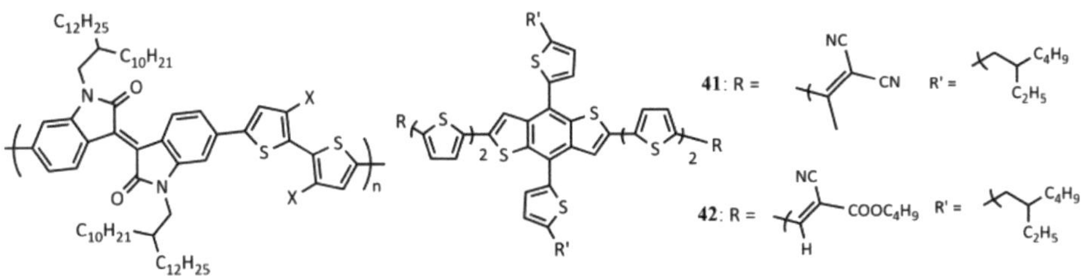

39-40

Fabrication of the solar cells was produced using *o*-xylene and diphenyl ether (DPE) as solvent and additive. To measure the photovoltaic performance of polymers the solar cells were fabricated using polymer sensitizers **39** and **40** with an inverted configuration (ITO/ZnO/polymer: PC71BM/MoO_3/Ag). The optimum blend ratio of polymer to PC71BM was 1:1.5 (w/w) for the two polymers. The polymer sensitizer **39** based cell showed a lower PCE of 4.92% with a V_{OC} of 0.89 V, a J_{SC} of 9.21 mA/cm^2, and a FF of 0.60. Whereas the sensitizer **40** exhibited a PCE of 8.80% with a V_{OC} of 1.06 V, a short-circuit current density (J_{SC}) of 12.58 mA/cm^2, and a FF of 0.66.

Novel dye sensitizers **41** and **42** with the sequence A-π-D-π-A which contains benzo[1,2-*b*:4,5-*b*']bisthiophene as a core moiety with different terminal acceptor were designed and synthesized. The effects of either methyl dicyanovinyl end group **41** or *n*-butyl cyanoester end group **42** on solubility, thermal properties, optical properties, charge transport, morphology, and photovoltaic performance were investigated [28]. Devices for these dye sensitizers **41** and **42** were fabricated at the optimal donor/acceptor weight ratio of 1:1 as-cast without annealing. Sensitizer **41** exhibited a short-circuit current (*J*sc) of 5.09 mA/cm^2, a V_{OC} of 1.09 V, a fill factor (FF) of 28.08%, and a PCE of 1.56% whereas, sensitizer **42** showed a *V*oc of 1.03 V and achieved a much better PCE performance of 6.17%, due to much higher FF of 59.08% and much higher short-circuit current (*J*sc) of 10.11 mA/cm^2. The external quantum efficiency (EQE) of these dyes have a similar broad photo response wavelength range of 300–700 nm while in the whole range, the EQE values of dye **42** are much higher than dye **41**. The EQE peak of **42** is about 48% at around 676 nm, while the EQE value of **41** is below 15% at all wavelength, which leads to the poor performance of the device. *J*sc values calculated from the EQE spectra are 3.36 mA/cm^2 for dye **41** and 9.89 mA/cm^2 for **42** respectively.

2.2 Fluorescent dyes

Fluorescence is a photophysical process which involves the emission of light by the substance as a consequence of the absorption of electromagnetic radiation. In most of the cases, the emitted light radiation has a longer wavelength (λ_{em}) than the absorbed light radiation (λ_{abs}). Likewise, fluorescent dyes, also known as *'fluorophores'* or *'reactive dyes'* remit light radiation upon absorption. Earlier, fluorescent dyes were extensively used in the textile industries to color fibers, cotton, yarns, and silk. Eventually, the use of fluorescent dyes has become a key technique for the detection and elucidation of biological structures by fluorescence emission technology. Because of their high photostability, and intense brightness, fluorescent dyes have been significantly used in fluorescent labeling (staining) of biomolecules. Fluorescent quenching studies have helped to detect DNA and proteins in biological systems. Techniques such as immunofluorescence, fluorescence microscopy, and flow cytometry rely upon fluorescent dyes. Currently, the requirement of fluorescent dyes insisted greatly because of their ample applications which could be substantiated through microwave-assisted synthesis. The advantages of microwave applications for the synthesis of fluorescent dyes have been intensively discussed [29].

2.2.1 Cyanine dyes

Cyanine dyes are found to be important functional dyes due to their typical optical properties, and act as sensitizers in solar cells, photography, and laser discs [30]. A significant property of cyanine dyes is the affinity for biological structures, specifically for DNA, and possesses wide color change, high photostability and increased fluorescent intensity when bound to biological structures [31]. Due to high fluorescence quantum yields and high molar extinction coefficients, they have been extensively used in cell imaging and gel staining techniques. Typically, cyanine dyes are obtained by heating a mixture of substituted quaternary salts with bisaldehyde or bis-imine. Accordingly, a series of cyanine dyes **43a-g** were synthesized by the condensation of quaternary salts of quinoline derivatives with 1*H*-indole-3-carbaldehydes in the presence of piperidine under solvent-free microwave irradiation at 126–329 W in 89–98% yields in only 2–5 min. The fluorescence spectra of the dyes showed absorption maxima (λ_{abs}) at 453–471 nm. However, in the presence of DNA, a bathochromic shift (red shift) at 483–499 nm was observed. Further, the living cell imaging experiments of the dyes **43b**, **43e**, and **44** have shown preferable staining of the head of the human sperm containing the nuclear DNA. Also, the motility of the sperm didn't slow down which indicated low cell cytotoxicity. Hence, these dyes could be used as potential fluorescent probes for labeling DNA to measure human sperm viability [31]. Likewise, the condensation of benzothiazole with quarternary salts of quinoline which upon coupling with the tertiary diamine linkers gave tetracationic analogous (bis-intercalators) of monomethine cyanine dyes **45**. The steady-state fluorescence spectral studies of **45** revealed greater labeling affinity toward DNA and proved for singlet oxygen sensitization property, and found to be a potential candidate for photodynamic therapy [32].

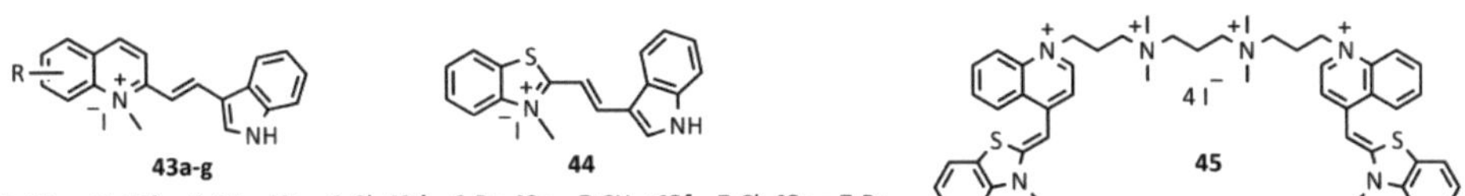

43a-g **44** **45**

R: **43a** = H; **43b** = 6-CH_3; **43c** = 6-Cl; **43d** = 6-Br; **43e** = 7-CH_3; **43f** = 7-Cl; **43g** = 7-Br

2.2.2 Naphthalimide dyes

1,8-Naphthalimide dyes are proved to be important fluorescent compounds due to their greater photostability and high fluorescent quantum yield. The basic spectral properties of these dyes depend on the polarization of naphthalimide molecule as a result of electron donor-acceptor interaction occurring between the substituents at the C-4 position and the carbonyl groups of the imide ring. Generally, 1,8-naphthalimide dyes are prepared *via* the substitution reaction of naphthalimides with various nucleophiles. The aromatic nucleophilic substitution reaction of 4-bromo-*N*-alkylnaphthalimides with amines, alkoxides, and thiols in the presence of KF/Al_2O_3 under solvent-free microwave irradiation yielded corresponding fluorescent dyes **46–48** which exhibited increased fluorescent intensity in the polar solvents [33]. Similarly, the derivatives **49** and **50** were obtained by the substitution reaction of 1,8-naphthalimides with primary amines. These dyes were further evaluated for the free radical scavenging properties against 2,2-diphenyl-1-picrylhydrazyl (DPPH). The results showed IC_{50} values at lower concentrations than the common synthetic antioxidant 2,6-ditertiarybutyl-4-methylphenol (BHT) [34].

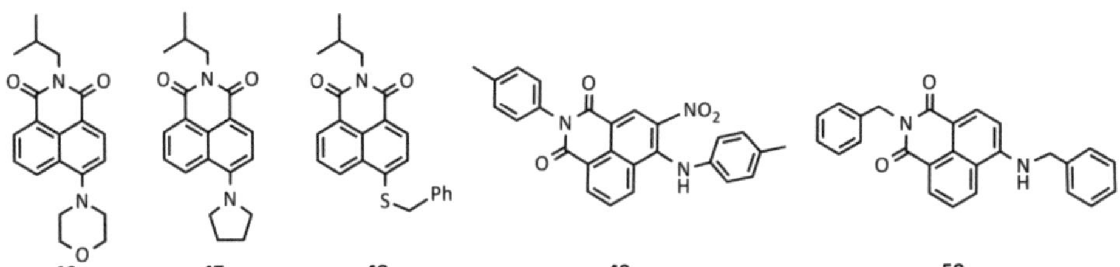

46 **47** **48** **49** **50**

2.2.3 Coumarin dyes

Coumarin dyes have been found commercial significance due to their intense fluorescence and are widely employed as fluorescent brighteners [35]. A one-pot microwave promoted synthesis of benzimidazol/benzoxazol functionalized coumarin dyes (**51a-e**) was developed which involved the reaction of 4-diethylam ino-2-hydroxybenzaldehyde, diethylmalonate, and *o*-phenylenediamine/ *o*-hydroxyaniline in *n*-pentanol within 3 min. The synthesized dyes showed higher fluorescence emission intensity and the dyes **51a-c** were further investigated for the effects of ink media on the fluorescence properties. Dye **51b** has exhibited an intense green fluorescence at 531 nm for mixed very long alkyl resin and maleic varnish in 60:40 ratios at 1% (w/w) concentration, and the fluorescence emission intensity of the dye reduced by 11% after 30 h of exposure to light. As a result, the dye **51b** could potentially be used in security offset ink [36]. Similarly, one-pot three-component microwave-assisted reaction of 7-diethyl-amino-coumarin ethylidene malononitrile, aromatic aldehydes and malononitrile to get highly fluorescent 3,5-disubstituted-2,6-dicyanoaniline coumarin dyes **52a-d** at 80°C in good yields in 2 min are prepared. The optical and thermal screening studies of **52a-d** exhibited excellent photophysical and thermal stability properties [37]. A group of 8-aza-7-hydroxy-4-methylcoumarin dye **53** was synthesized by reacting 2,6-dihydroxypyridine hydrochloride with ethylacetoacetate in the presence of magnesium bromide as a Lewis acid catalyst. This fluorophore is adequately soluble in water and has a high fluorescent quantum yield and showed increased fluorescence in protic solvents at neutral pH, which could be useful in biosensors that are required for finding biologically active compounds [38]. Furthermore, a microwave-assisted Knoevenagel condensation of salicylaldehyde and cyano-N′-methyleneacetohydrazide in the presence of piperidine catalyst gave 3-carbohydrazide coumarin fluorescent dye (**54**) which could be used to print polyester and polyamide fabrics [35].

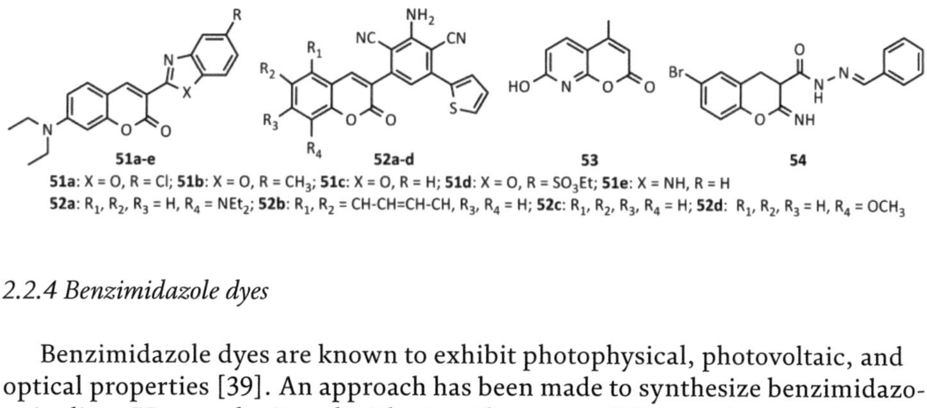

51a-e **52a-d** **53** **54**

51a: X = O, R = Cl; **51b**: X = O, R = CH_3; **51c**: X = O, R = H; **51d**: X = O, R = SO_3Et; **51e**: X = NH, R = H
52a: R_1, R_2, R_3 = H, R_4 = NEt_2; **52b**: R_1, R_2 = CH-CH=CH-CH, R_3, R_4 = H; **52c**: R_1, R_2, R_3, R_4 = H; **52d**: R_1, R_2, R_3 = H, R_4 = OCH_3

2.2.4 Benzimidazole dyes

Benzimidazole dyes are known to exhibit photophysical, photovoltaic, and optical properties [39]. An approach has been made to synthesize benzimidazoquinolines **55a-c**, substituted with piperidine, pyrrolidine, and piperazine moieties by uncatalyzed amination protocol under microwave heating in relatively high yields (56–90%), which by conventional heating after several days gave **55a-c** only in low yields (<10%). The emission spectra of **55a-c** showed an increase in the fluorescence intensity when interacted with the calf thymus DNA (*ct*-DNA) [40]. The microwave promoted synthesis of bis-benzimidazolyl derivatives upon *N*-alkylation gave water-soluble fluorescent dyes **56a-b**. These dyes proved to be highly selective fluorescent probe toward Zn^{2+} in aqueous solution and the mixture of dye-Zn^{2+} could detect picric acid by fluorescence quenching [41]. Under solvent-free microwave irradiation, a series of 2-substituted styryl benzimidazole dyes **57a-g** and **58a-f** were prepared by the condensation of 2-alkyl benzimidazoles with aromatic aldehydes in the presence of acetic anhydride [42].

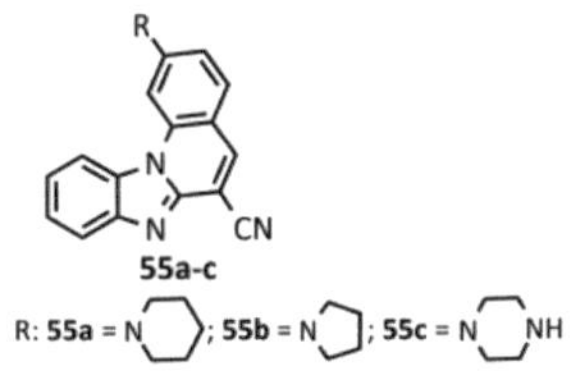

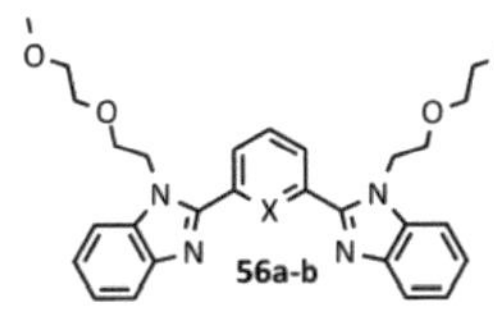

56a-b

X: 56a = N; 56b = H

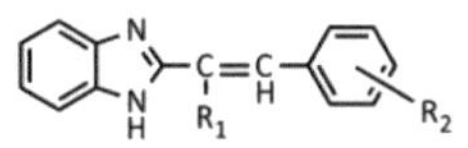

57a-g, 58a-f

57: R_1= H, R_2: **57a** = *p*-OCH_3; **57b** = *p*-OH; **57c** = H; **57d** = *p*-Cl; **57e** = *p*-CH_3; **57f** = *m*-NO_2; **57g** = *p*-$N(CH_3)_2$;
58: R_1= CH_3, R_2: **58a** = *p*-OCH_3; **58b** = *p*-OH; **58c** = H; **58d** = *p*-Cl; **58e** = *p*-CH_3; **58f** = *m*-NO_2

2.2.5 Imidazole dyes

The imidazole moiety is immensely employed in DSSC's [43]. Interestingly these dyes **59a-d**, **60**, and **61** are prepared by one-pot condensation of α-diketone (benzil), aryl aldehydes, and ammonium acetate in the presence of glacial acetic acid under microwave irradiation. Furthermore, the dyes **59a-d** have been proved to be potential antimicrobial agents against *E. coli*, *B. subtilis*, *S. aureus,* and *L. monocytogenes* [44]. The dye **60** exhibited a strong two-photon upconverted blue fluorescent emission peak around 443–476 nm [45].

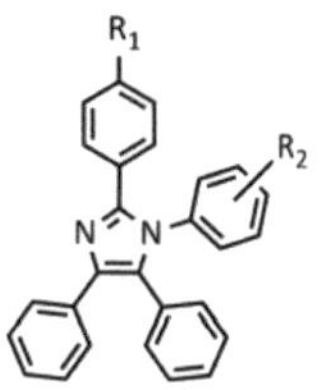

59a-d

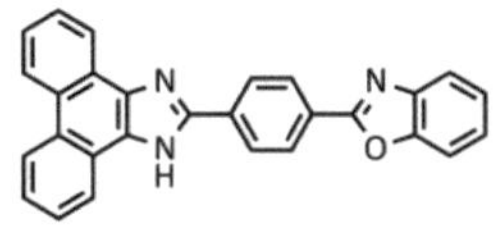

60

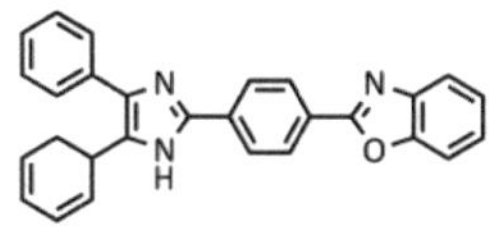

61

59a: R_1 = NO_2, R_2 = 4-OCH_3; **59b**: R_1 = COOH, R_2 = 4-OCH_3; **59c**: R_1 = COOH, R_2 = 4-COOH; **59d**: R_1 = COOH, R_2 = 2,5-COOH

2.2.6 Thiophene dyes

Thiophene oligomers and polymers have put forward extensive applications in organic electronics, owing to their remarkable performance as organic semiconductors [46]. A series of thiophene oligomer based fluorophores appended with 4-sulfo-2,3,5,6-tetraflurophenyl ester **62a**, *N*-hydroxysuccinimidyl ester **62b**, and phthalimide **63a-b** are prepared efficiently in shorter reaction times by sequential Pd(II) catalyzed Suzuki cross coupling reaction by taking advantage of microwave irradiation. The dyes **62a-b** were evaluated for their labeling toward monoclonal antibodies Anti-CD38. The dye **62a** showed a larger bathochromic shift compared to **62b** and exhibited greater affinity toward the monoclonal antibody [47]. The cyclic voltammetry, UV–visible spectroscopy, and X-ray crystallographic studies of the dyes **63a** and **63b** revealed π-π stacking packing mode which led to increased charge carrier mobility envisaging as an ambipolar semiconductor with applications in both Organic Thin-Film transistors (OTFT) and Organic-light Emitting Transistors (OLET) [46, 48]. A one-pot three-component synthetic route was used to prepare thiophene-coumarin based dyes **64a-j** in 92–96% yields from hours to min by the use of microwave irradiation technique from 3-acetyl coumarin, malononitrile, and elemental sulphur (S_8). The spectroscopic data of the dyes **64a-j** showed a bathochromic shift in various solvents. The dye **64g** was further investigated for its pH sensitivity *via* deprotonation and reverse protonation in two solvent systems (DMSO and DMSO/H_2O binary mixture) using absorption and fluorescence techniques. The -OH group of **64g** is susceptible to deprotonation under alkaline medium (TBAOH, tetrabutylammonium hydroxide) and reverse protonation by the addition of trifluoroacetic acid (TFA). A distinct fluorescence color change from light blue to green was observed with the incremental addition of TBAOH to the solution of **64g** and reverse phenomena was observed with the incremental addition of TFA [49].

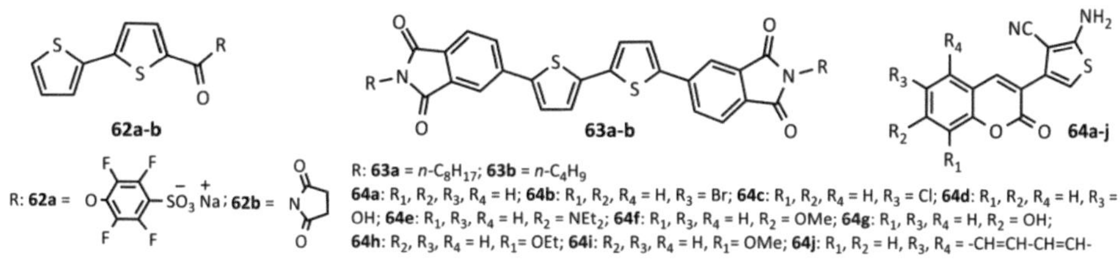

2.2.7 Inorganic dyes

Inorganic dyes are procured when the organic dyes are combined with appropriate metals. Typically monoazodyes containing additional groups such as amino, hydroxyl, and carboxyl groups which are capable of forming coordination complexes with metal ions are used. This organo-metallic combination could lead to enhanced optical properties. The synthesis of organo soluble 4-*t*-butylphthalocyanine (TBPc) and organo soluble sodium salt of sulfonated phthalocyanine (Pc-SO_3Na) metal complexes of Cu^{2+}, Mg^{2+}, and Zn^{2+} (**65a-b**) has been reported. Further, lutetium complex [Lu(TBPor)(TBPc)] **66** ligated with 4-*t*-butylporphyrin (TBpor) and 4-*t*-butylphthalocyanine (TBPc) rings were obtained *via* the reaction of lutetium acetate (LuOAc) with corresponding ligands under microwave irradiation. The prepared complexes were blended with *N*,*N′*-bis-(1,5-dimethylhexyl)-3,4:9,10-perylene-bis-(dicarboximide) [PDHEP] and SnO_2 glass to fabricate photoelectric cells. The SnO_2 glass/Mg-Pc$(SO_3Na)_4$/PDHEP/Al photoelectric cell exhibited a short-circuit photocurrent of 116 μA/cm^2, whereas SnO_2 glass/Lu(TBpor)(TBpc)/PDHEP/TiO_2/Al photoelectric cell showed increased short-circuit photocurrent of 691.3 l μA/cm^2 under the illumination of white light at 1.201 mW/cm^2 [50]. The metal-free phthalocyanine and metallophthalocyanine complexes (**67** and **68a-c**) of Cu^+, Cu^{2+}, Co^{2+}, Ni^{2+}, Fe^{2+}, Zn^{2+}, Pd^{2+}, Pt^{4+}, and Ru^{3+} was prepared by the reaction of corresponding azo dyes with metal salts using microwave heating, which were obtained in poor yields by conventional heating [51].

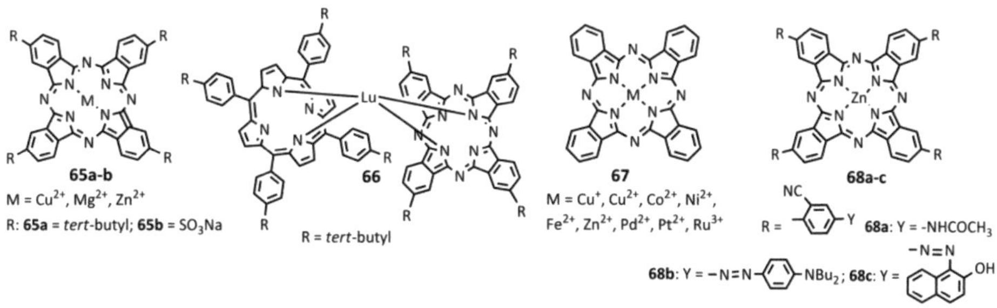

2.3 Photochromatic dyes

Some materials at their molecular level exhibit a property of changing their absorption spectra on exposure to light radiation. This is usually a reversible change and is accompanied with alteration in the physical or chemical property. This kind of photo transformation is referred to as photochromism. The reverse change may be induced thermally (photochromism type T) or photochemically (phtochromism type P). The discovery of photochromic materials can be retraced to the middle of 19th century when Hirshberg and his team (1950) have contributed significantly towards the synthesis and mechanistic studies of photochromic materials. Hirshberg coined the term "Photochromism" from Greek words 'photos' meaning light and 'chroma' means color. Varieties of materials like minerals, nanoparticles, inorganic–organic compounds, organic dyes, polymers, and biomolecules have been explored to exhibit photochromic property. They have been in use in modern applications like erasable optical memory media, photo-optical switch components, sunscreen applications, contact lenses, security glasses, and thin films. Some of the organic photochomic compounds undergo reversible light-driven reaction hence these compounds are often incorporated into polymers, liquid crystals, and other such matrices. Although the decade 1950–1960 has remarked synthesis of photochromic materials with the advancement in newer supportive technologies such as spectroscopy the field has not gained acceleration. This is due to the sensibility of organic materials towards the light which makes them undergo degradation (they were not fatigue resistant). After the report of the synthesis of fatigue resistant spironapthoxazines many-fold increase in the applications of photochromic materials has been reported. Spiropyrans, spriooxazines, chromenes, fulgides, fulgimides, diarylethenes, spirodihydro indazolines, azocompounds, polyarenes, quinones, anils are the photochromic dyes in the industrial and general application field [52]. In the recent past attempts have been made to apply microwave-assisted synthetic methods to the total synthesis or in one or two intermediate steps.

Spirooxazines are the important photochromic dyes being popularly seen in very common to high tech applications. Due to their brilliant light fatigue resistance nature, they are the dyes of bright prospects. The reports of the synthesis of spirooxazines by conventional methods are many. Successful efforts have also been made to obtain them by environment-friendly microwave-assisted synthetic methods. Spirooxazine **69a** has been prepared under microwave conditions, starting from 1-nitroso-2-naphthol in presence of triethylamine as a catalyst in a CEM focused microwave reactor provided with temperature control [53]. Electric power 25–35 W, temperature 80–180°C were found to be the optimized conditions to get yields comparable to the traditional thermal method. Indolinespironaphthooxazine **69b-d** have been prepared from 1-nitroso-2,7 dihydroxy naphthalene by a microwave irradiation technique [54]. The reaction was carried out in microwave synthesizer (MAS-I). Microwave irradiation was done at 600 W. The products were obtained in very good yield within a few minutes of reaction time.

Fulgides **70a-b** and their derivatives fulgimides **71a-b** are an important class of photochromatic materials used mainly in optical memory devices and optical switches. Fulgides are intense colored compounds which are good in resisting the photodegredation in comparison to fulgimides. However, fulgimides have better resistance to acid or base hydrolysis further that their N-substituent can be used as a link to prepare photochromic films. Both these classes of compounds have been thoroughly researched. A successful attempt to synthesize fulgimides using domestic microwave ovens has been made [55]. As compared to classical thermal method microwave-assisted synthesis has led to 3 fold times reduction in duration of synthesis, an increase in the yield up to 2 times, and minimization of the use of organic solvents. The efficient synthesis of N-functionalized fulgimides **72** was achieved under microwave irradiation [56]. Fulgides were converted to fulgimides in two steps in the presence of DMAP and DCC by microwave irradiation in presence of pyridine and xylene as the solvent. They have attained from 50 to 84% increase in yield by benign microwave method in very short reaction time. Oxazole and indole based heterocyclic fulgides **73, 74** were synthesized by microwave method using clay as a catalyst from fulgenic acids [57]. Their synthesis involved stirring of the blended mixture of fulgenic acid and montmorillonite KSF along with isopropyl acetate in a flask. The yield was improved to 72–84% by MWAS as compared to the conventional Stobbe condensation method. One-pot three-component microwave-assisted synthesis of novel azo-imidazoles **75a-h** is reported which exhibited photochromatic property with UV–Visible light [58]. Azo dye, ammoniumacetate, and benzil were reacted under microwave irradiation using acetic acid as solvent.

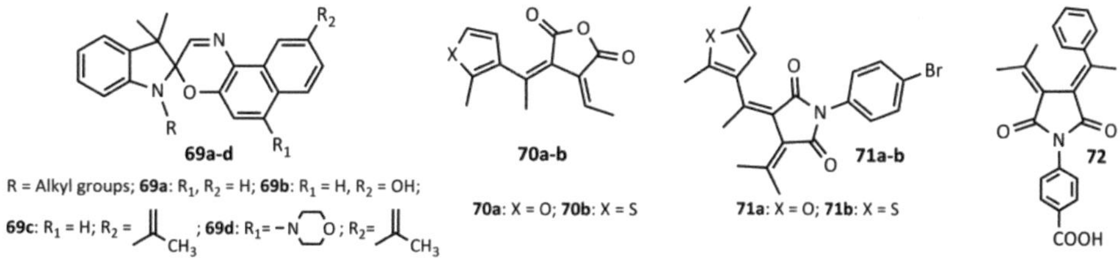

At optimal power 230 W microwave irradiation for 2 min duration 87% yield of the dye **75a-h** was obtained. It did not involve any thermal degradation by-products and economical use of organic solvents makes this protocol a green synthetic method. The microwave synthetic method was applied to successfully prepare photochromic spiropyran **76** [59]. Spiropyrans are spiro-fused indolochromenes. Due to their photochromic isomerization property, they are used in optical switches and sensors. The synthesis involves one-pot two-step reaction. Initially, water-mediated reaction was carried out between 1, 2, 3-trimethylindole, and benzyl bromide under microwave environment. Microwave irradiation was done at De Rosa and Soriente's conditions (i.e. 200 MW power and 150°C temperature for 8 minutes). Then, the resulting reaction mixture after a simple workup procedure was treated with 5-nitrosalicylaldehyde under microwave irradiation using ethanol as the solvent. They have obtained product **76** in excellent yield after the flash chromatographic workup procedure. It is an environmentally benign synthetic method using a minimum amount of solvent.

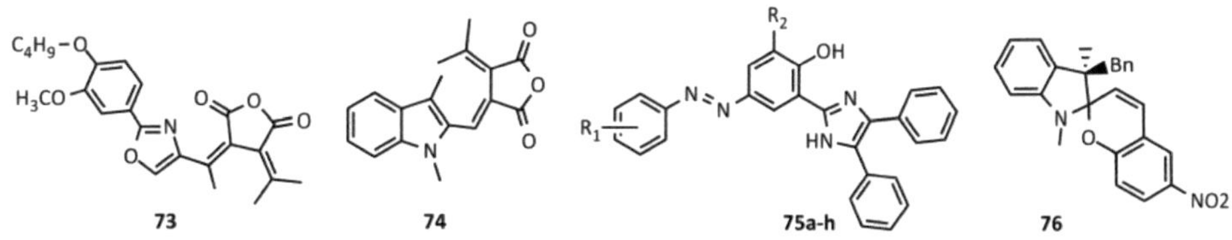

75a: R_1, R_2 = H; **75b**: R_1 = 4-CH_3, R_2 = H; **75c**: R_1 = 4-Cl, R_2 = H; **75d**: R_1 = H, R_2 = OCH_3; **75e**: R_1 = 3-NO_2, R_2 = OCH_3; **75f**: R_1 = 2-CH_3, R_2 = OCH_3; **75g**: R_1 = 4-NO_2, R_2 = OCH_3; **75h**: R_1 = 4-F, R_2 = OCH_3

2.4 Organic-light emitting diodes (*O*-LEDs)

The light-emitting diode (LED) is a light-emitting semiconducting material when current flows through it. The current flow induced light emission was first observed by Captain Henry Joseph Round in 1907. Light emission takes place when electrons undergo a transition from the conduction band to the empty valence band. The band gap in semiconducting material decides the color of emitted light. *O*-LED are the LEDs in which the light-emissive electroluminescent film is made up of organic molecules. In the case of *O*-LEDs the highest occupied molecular orbital (HOMO) is the conduction band, and the lowest unoccupied molecular orbital (LUMO) is the empty valence band of organic substance. *O*-LEDs are preferred over LEDs due to the facts that an *O*-LED is thinner and have a better display property; it has brighter, fast responsive, and long-range contrast display. Moreover, *O*-LEDs have wider viewing angles with low driving voltage property. *O*-LEDs can be conveniently fabricated on a glass surface at low temperatures. Organic semiconducting materials are in the crystalline or polymeric phase. Organometallic compounds, polymers, and even simple organic molecules like aryl amines are used in *O*-LEDs. The research in the field of *O*-LEDs is in rapid progress as these displays are already in use in modern electronic and optoelectronic appliances like heads-up displays, billboard-type displays, automotive dashboards, home and office lighting, and flexible displays. The synthetic invention of these organic moieties is a progressive field, and the microwave-assisted synthetic methods of *O*-LEDs have also started sprouting in recent years [60, 61].

The amalgamation of organic moieties and inorganic matrices results in the synergetic effects by augmenting of the properties like flexibility and shape ability with stability [62]. Poly (2-hydroxyethyl methacrylate) (PHEMA) silica-hybrids have been prepared by microwave irradiation [63]. Organoboron dye diketonate BF_2 complex **77**, borondipyrromethene (BODIPY) **78**, and (1,3-boron di(iso)indomethene dye **79** can be integrated into these PHEMA silica hybrids.

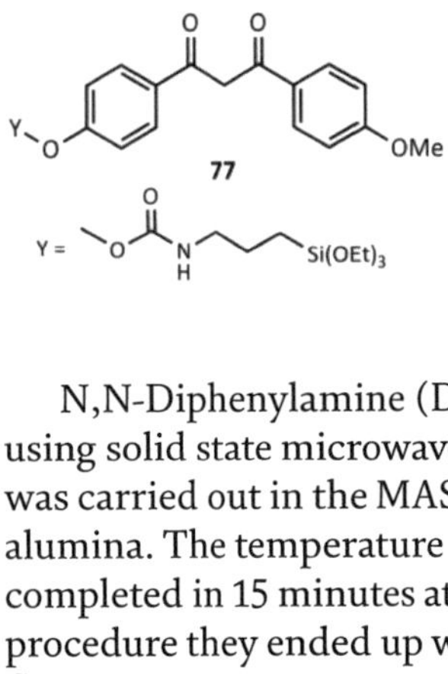

77

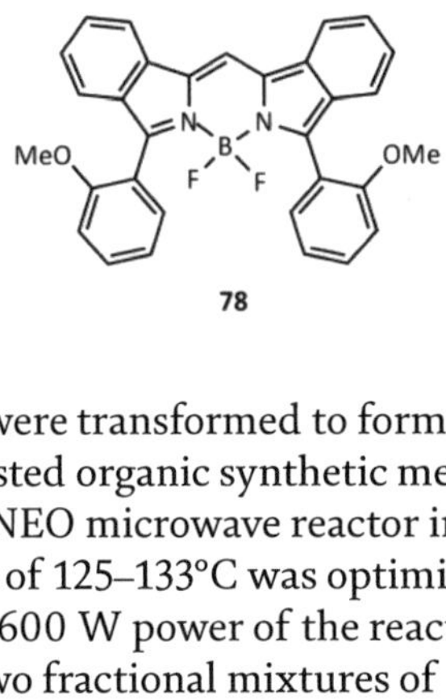

78

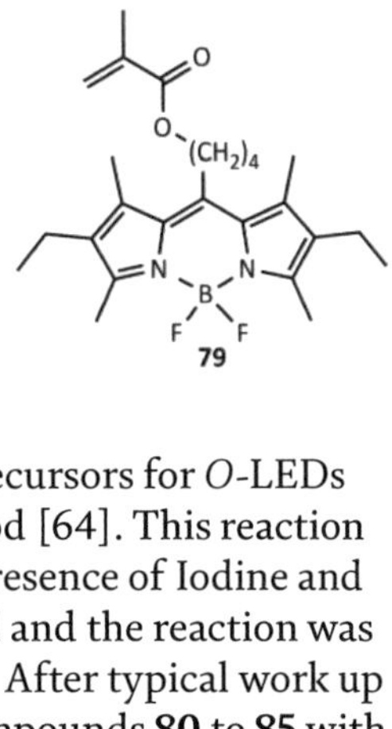

79

N,N-Diphenylamine (DPA) were transformed to form precursors for *O*-LEDs using solid state microwave-assisted organic synthetic method [64]. This reaction was carried out in the MAS II SINEO microwave reactor in presence of Iodine and alumina. The temperature range of 125–133°C was optimized and the reaction was completed in 15 minutes at 500–600 W power of the reactor. After typical work up procedure they ended up with two fractional mixtures of compounds **80** to **85** with fluorescence property were obtained.

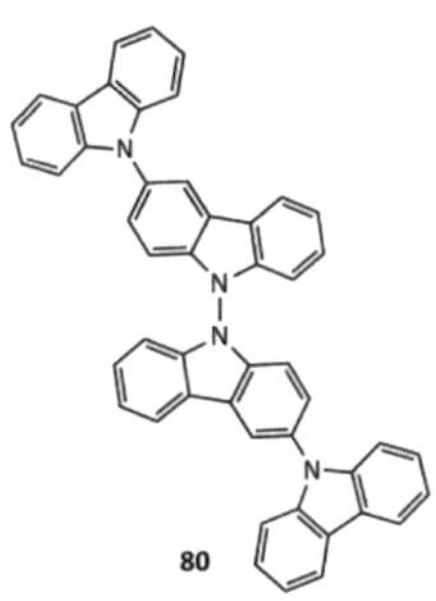

80

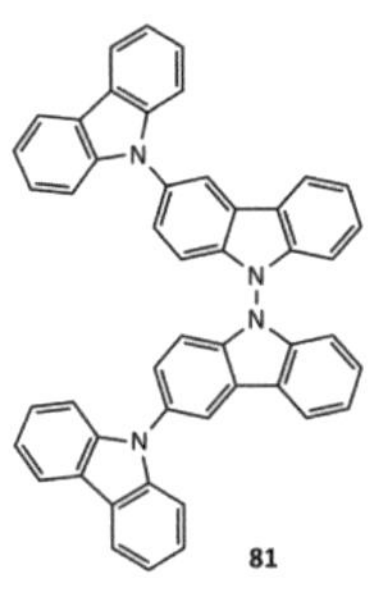

81

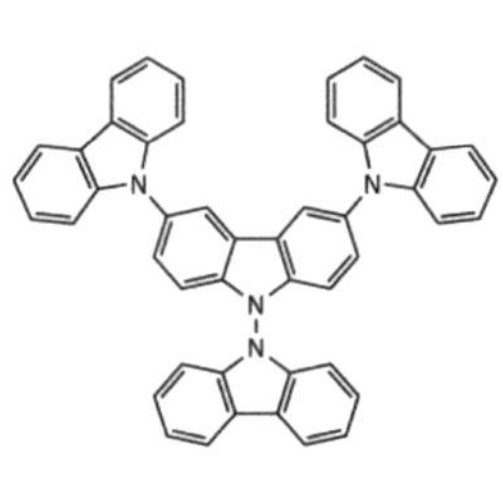

82

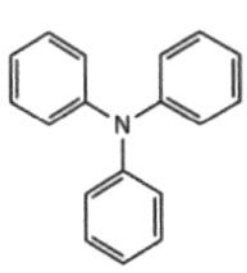

83

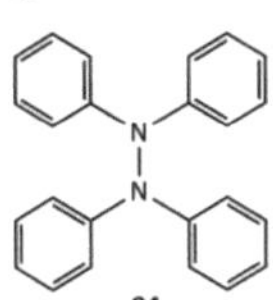

84

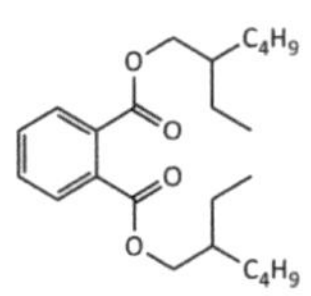

85

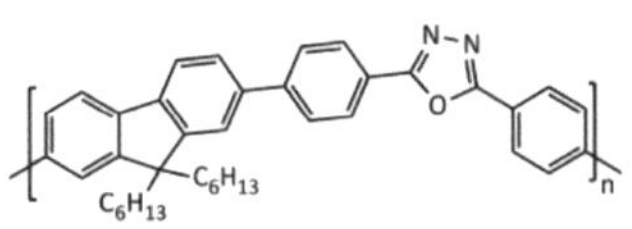

86

DOI: http://dx.doi.org/10.5772/intechopen.94946

Polyfluorene is regarded as an important source for the development of *O*-LEDs. It emits blue light and the color of the light can be tuned by means of doping, structural engineering, preparing materials with tuned properties [65]. Microwave-assisted synthesis of π conjugated polymers were reported and this method was proved to be an advantageous method over multi-step expensive conventional method. They have effectively used microwave conditions in oxidative polymerization of 2,5-diphenyl-1,3,4-oxadiazole and 9,9-dihexyl-fluorene monomers to get poly (dihexyl fluorene-co diphenyl oxadiazole) (POF) **86** in the presence of $FeCl_3$ catalyst.

3. Conclusions

Since from the centuries, dyes have played a very important role in human life. The functional dyes have changed the technologies drastically and have gained immense importance now a day. A specific property of the dye depends on the various factors such as the donor, electron acceptor/π-conjugation, linker, etc. present at appropriate positions. More effort has been established into searching for better dyes with expected properties. Microwave-assisted synthesis has changed the methodology of organic synthesis and hence is also efficiently applied in the synthesis of functional dyes. Therefore, a number of dyes synthesized under microwaves along with their applications were discussed. There is a possibility for further development in organic synthetic methodology under microwaves to obtain dyes having wider applications in organic photovoltaics, fluorescence sensors, photochromic materials, OLEDs, etc.

Acknowledgements

The authors thank the DST, New Delhi for the sanction of PURSE Phase -II to the Karnatak University, Dharwad.

Author details

Sheetal Marganakop[1], Pramod Kattimani[2], Sudha Belgur Satyanarayana[3] and Ravindra Kamble[4*]

1 Government First Grade College, Paschapur, Karnataka, India

2 J.S.S. Arts, Science and Commerce College, Falls Road, Gokak, India

3 Department of Chemistry, Yuvaraja's College, University of Mysore, Mysore, India

4 Department of Chemistry, Karnatak University, Dharwad, India

*Address all correspondence to: ravichem@kud.ac.in

References

[1] Lauf RJ, Bible DW, Johnson AC, Everliegh CA. 2-18 GHz broadband microwave heating systems. Microwave Journal 1993;36:24-27.

[2] (a) Lidstrom P, Tierney J, Wathey B, Westman J. Microwave-assisted organic synthesis- a review. Tetrahedron. 2001; 57:9225-9283. DOI: org/10.1016/ S0040-4020 (01)00906-1 (*b*) Roberts BA, Strauss CR. Towards rapid, "green", predictable microwave-assisted synthesis. Accounts of Chemical Research. 2005;38:653-661. DOI: org/10.1021/ar040278m; (c) Wiesbrock F, Hoogenboom R, Schubert US. Microwave-assisted polymer synthesis: State-of-the-art and future perspectives. Macromolecular Rapid Communications. 2004;25:1739-1764. DOI:org/10.1002/marc.200400313 (d) Kappe CO, Dallinger D. The impact of microwave synthesis on drug discovery. Nature Reviews Drug Discovery. 2006;5:51-63. DOI: 10.1038/ nrd1926; (e) Wenbin Cao, editor. The Development and Applications of Microwave heating. London: Intech Open; 2012. DOI: 10.5772/2619.

[3] Gronnow MJ, White RJ, Clark JH, Macquarrie DJ. Energy efficiency in Chemical Reactions: A comparative study of different reaction techniques. Organic Process Research Development. 2005;9:516-518. DOI: org/10.1021/ op0498060.

[4] Thostenson ET, Chou TW. Microwave processing: fundamentals and applications, Composites: Part A. 1999;30:1055-1071. DOI: org/10.1016/ S1359-835X(99)00020-2.

[5] Ferreira ES, Hulme AN, McNab H, Quye A. The natural constituents of historical textile dyes. Chemical Society Reviews. 2004;33:329-336. DOI: org/10.1039/B305697J.

[6] Sivakumar V, Lakshmi AJ, Vijayeeshwari J, Swaminathan G. Ultrasound assisted enhancement in natural dye extraction from beetroot for industrial applications and natural dyeing of leather. Ultrasonics Sonochemistry. 2009;16:782-789. DOI: org/10.1016/j.ultsonch.2009.03.009.

[7] Cova, TFGG, Pais AACC, Seixas de Melo JS. Reconstructing the historical synthesis of mauveine from Perkin and Caro: procedure and details. Science Reports. 2017;7: 6806. DOI: org/10.1038/ s41598-017-07239-z.

[8] Yoshida Z, Kitao T. Chemistry of Functional Dyes. Tokyo: Mita Press; 1989

[9] (a) Armaroli N, Balzani V. The Future of Energy Supply: Challenges and Opportunities. Angewandte Chemie International Edition. 2007;46:52-66; (b) Gust D, Moore TA, Moore AL. Solar Fuels via Artificial Photosynthesis. Accounts of Chemical Research. 2009;42:1890-1898 (*c*) Nocera DG. Chemistry of Personalized Solar Energy. Inorganic Chemistry. 2009; 48:10001-10017.

[10] (a) O Regan B, Gratzel M. A low-cost, high-efficiency solar cell based on dye-sensitized colloidal TiO_2 films. Nature 1991;353:737-740. DOI: org/10.1038/353737a0; (b) Gratzel M. Perspectives for dye-sensitized nanocrystalline solar cells. Progress in Photovoltaics. 2000;8:171-185. DOI:10.1002/ (SICI)1099-159X(200001/02)8.

[11] (a) Gratzel M. Photoelectrochemical cells. Nature 2001;414:338-344. DOI: org/10.1038/35104607; (b) Guunes S, Sariciftci NS. Hybrid solar cells. Inorganica Chimica Acta. 2008;361:581-588. DOI: org/10.1016/j. ica.2007.06.042.

[12] (a) Dallinger DC, Kappe O. Microwave-Assisted Synthesis in Water as Solvent, Chemical Reviews.

2007;107:2563-2591. DOI: org/10.1021/cr0509410; (b) Sharma N, Sharma UK, Van der Eycken EV. Microwave-Assisted Organic Synthesis: Overview of Recent Applications. In: Zhang W, Cue BW, editors. Green techniques for organic synthesis and medicinal chemistry. Chichester: Wiley; 2012. 441 p. DOI: org/10.1002/9781119288152.ch17; (c) Bogdal D. Microwave-assisted Organic Synthesis: One hundred Reactions, Volume 25, 1st ed. Elsevier Science; 2005.p.214 p.

[13] Egor VV, Aleksandr VS, Nadezhda IM, Igor VD, Anatoly VM, Vladimir IM, Sergey AK, Victor VE, Vitaly AG, Oleg NC, Gennady LR, Valery NC. Synthesis, Photophysical and redox properties of the D-π-A type pyrimidine dyes bearing the 9-phenyl-9*H*-carbazole moiety. Journal of Fluorescence. 2015;25:763-775, DOI: 10.1007/s10895-015-1565-6.

[14] Lijie L, Lipeng Z, Miao L, Yijing G, Jinsheng S, Hua W. Dyes and Pigments. 2016;130:63-69. DOI: 10.1016/j.dyepig.2016.03.026.

[15] Mohammad NS, Muhammad MU, Nisar U, Abdulrahman FA, Abdulaziz AS, Dyes and Pigments, 2017;141:406-412. DOI: org/10.1016/j.dyepig.2017.02.041.

[16] Alberto F, Roberta M, Andrea M, Maria FO, Alberto M, Fabio M, Stefania S. Dyes and Pigments. 2017;145:246-255. DOI: 10.1016/j.dyepig.2017.05.058.

[17] Pastore M, De Angelis F. Aggregation of organic dyes on TiO_2 in dye-sensitized solar cells models: an *ab initio* investigation. ACS Nano. 2010;4:556-62. DOI: org/10.1021/nn901518s.

[18] Xiaodong C, Junfeng T, Zhicai H, Mingjing Z, Xiaofang Z, Jun M, Peili G, Jianfeng L, Peng Z, Chenglong W, Yangjun X, Hongbin W. Dyes and Pigments. 2018;158:319-325. DOI: 10.1016/j.dyepig.2018.05.052.

[19] Carlos AE, Alberto I, María ÁH, Alejandro O, Robert C, Vivek D, Luis E, Braulio I, Nazario M. Dyes and Pigments. 2014;107:9-14. DOI: 10.1016/j.dyepig.2014.03.010.

[20] (a) Roncali J, Blanchard P, Leriche P. Molecular materials for organic photovoltaics: Small is beautiful. Advanced Materials 2014;26:3821-3838. DOI: org/10.1002/adma.201305999; (b) Lin Y, Li Y, Zhan X. Small molecule semiconductors for high-efficiency organic photovoltaics. Chemical Society Reviews. 2012;41:4245-4272. DOI: org/10.1039/C2CS15313K.

[21] Agostina LC, Luisa DM, Eduardo F, Roberto G, Angela S, Claudia C, Giuseppina AC, Maria PC, Giuseppe G, Giuseppe C. Journal of Materials Chemistry - A. 2014;2:14181-14188. DOI: org/10.1039/C4TA02161D.

[22] Agostina LC, Luisa DM, Giuseppina AC, Roberto G, Eduardo F, Antonio C, Giuseppe G, Giuseppe C. Dyes and Pigments. 2016;130:79-89. DOI: org/10.1016/j.dyepig.2016.02.030.

[23] Mauricio C, Carlos AE, Robson RG, Alejandro O, Koiti A, Braulio I. Journal of Molecular Structure. 2017;1133:384-391. DOI: org/10.1016/j.molstruc.2016.12.021.

[24] Yogesh ST, Chaochin S, Ming-Tai S, Sheng-Han T, Shih-Yu H, Wen-Ren L. New Oxindole-Bridged Acceptors for Organic Sensitizers: Substitution and Performance Studies in Dye-Sensitized Solar Cells. Molecules 2020;25:2159. DOI: 10.3390/molecules25092159.

[25] Chen, BS, Chen DY, Chen, CL, Hsu, CW, Hsu HC, Wu KL, Liu SH, Chou PT, Chi Y. Donor-acceptor dyes with fluorine substituted phenylene spacer for dye-sensitized solar cells. Journal of Materials Chemistry. 2011;21:1937-1945. DOI: org/10.1039/C0JM02433C.

[26] Alessio D, Massimo C, Alessandro M, Maurizio P, Adalgisa S, Riccardo B, Fabrizia FB, Maurizio T, Daniele C, Aldo di C, Gianna R, Lorenzo Z. RSC Advances. 2015;5: 32657-32668. DOI: org/10.1039/C5RA03530A.

[27] Eui HJ, Hyungju A, Won HJ, Jea WJ, Jae WJ. Isoindigo-based conjugated polymer for high-performance organic solar cell with a high V_{OC} of 1.06 V as processed from non-halogenated solvent. Dyes and Pigments. 2019;161:113-118. DOI: org/10.1016/j.dyepig.2018.09.048.

[28] Jie G, Dmitry OB, Chengjun G, Svetlana MP, Sergei AP, Zhitian L, Yurri NL, Jie M, Aiwen L. End group tuning in small molecule donors for non-fullerene organic solar cells. Dyes and Pigments. 2020;175:108078.DOI: org/10.1016/j.dyepig.2019.108078.

[29] Galal HE, Reham AM. Microwave synthesis of fluorescent and luminescent dyes (1990-2017). Journal of Molecular Structure. 2018;1173:707-742. DOI: 10.1016/j.molstruc.2018.06.101.

[30] (a) Masaki M, Yoshimi H, Kazumasa F, Ji-Yen J, Tsukasa Y, Hideki M. Application of near-infrared absorbing heptamethine cyanine dyes as sensitizers for zinc oxide solar cell. Synthetic metals. 2005;148:147-153. DOI: 10.1016/j.synthmet.2004.09.026; (b) Karatsu T, Karatsu T, Yanai M, Yagai S, Mizukami J, Urano T, Kitamura A. Evaluation of sensitizing ability of barbiturate-functionalized non-ionic cyanine dyes; application for photoinduced radical generation system initiated by near IR light. Journal of Photochemistry and Photobiology A: Chemistry. 2005;170:123-129. DOI: 10.1016/j.jphotochem.2004.08.010; (c) Raz G, Gary R, Paras NP. The influence of structure and environment on spectroscopic and lasing properties of dye-doped glasses. Optical Materials. 1997;8:43-54. DOI: 10.1016/S0925-3467(97)00035-9.

[31] (a) Xiang-Han Z, Lan-Ying W, Zhi-Xiang N, Shi-Huan T, Zu-Xun Z. Microwave-assisted solvent-free synthesis and spectral properties of some dimethine cyanine dyes as fluorescent dyes for DNA detection. Dyes and Pigments. 2008;79:205-209. DOI: 10.1016/j.dyepig.2008.02.010; (b) Zhang XH, Liu Q, Shi HJ, Wang LY, Fu YL, Wei XC, Yang LF. Synthesis, spectral properties of cell-permeant dimethine cyanine dyes and their application as fluorescent probes in living cell imaging and flowcytometry. Dyes and Pigments. 2014;100:232-240. DOI: 10.1016/j.dyepig.2013.09.011.

[32] Hussein HA, Wael AES, Mohamed HA, Mahasen SA, El-Zeiny ME. Microwave synthesis and fluorescence properties of homo- and heterodimeric monomethine cyanine dyes TOTO and their precursors. Green Chemistry Letters and Reviews. 2016;10:10-22. DOI: 10.1080/17518253.2016.1258088.

[33] Bardajee GR. Microwave-assisted solvent-free synthesis of fluorescent naphthalimide dyes. Dyes and Pigments. 2013;99:52-58. DOI: 10.1016/j.dyepig.2013.04.004.

[34] Ye Z, Shaobo F, Qiang W, Kai W, Xianghui Y, Hengshan W, Yingming P. Microwave-assisted synthesis and evaluation of naphthalimides derivatives as free radical scavengers. Medicinal Chemistry Research. 2011;20:752-759. DOI 10.1007/s00044-010-9384-4.

[35] Elgemeie GH, Ahmed KA, Ahmed EA, Helal MH, Masoud DM. A simple approach for the synthesis of coumarin fluorescent dyes under microwave irradiation and their application in textile printing. Pigment & Resin Technology. 2015;45:217-224. DOI: 10.1108/PRT-02-2015-0019.

[36] (a) Farahnaz N, Mahnaz DG. Microwave-Promoted One-Pot Syntheses of Coumarin Dyes. Synthetic Communications. 2010;40:901-909. DOI: 10.1080/00397910903026699; (b) Fatemeh T, Farahnaz N, Saeed B. Development of novel fluorescent offset ink based on coumarin dyes: Synthesis and properties. Progress in Organic Coatings. 2014;77:1351-1359. DOI: 10.1016/j.porgcoat.2014.04.022.

[37] Burcu A, Ergin Y, Vildan K, Zeynel S. Efficient one-pot three-component method for the synthesis of highly fluorescent coumarin-containing 3,5-disubstituted-2,6-dicyanoaniline derivatives under microwave irradiation. Synthetic Communications. 2017;23:2174-2188. DOI: 10.1080/00397911.2017.1362438.

[38] Hikaru T, Tetsuo N, Wataru N, Hirokazu T. Microwave-Assisted Synthesis of Azacoumarin Fluorophores and the Fluorescence Characterization. Journal of Organic Chemistry. 2017;82: 2739-2744. DOI: 10.1021/acs.joc.6b02656.

[39] Govardhana BB, Justin Thomas KR, Miao-Syuan F, Kuo-Chuan H. Benzimidazole-Branched Isomeric Dyes: Effect of Molecular Constitution on Photophysical, Electrochemical, and Photovoltaic Properties, Journal of Organic Chemistry. 2016;81:640-653, DOI: 10.1021/acs.joc.5b02590.

[40] Natasa P, Marijana H, Gordana P, Grace KZ. Novel aminated benzimidazo[1,2-*a*]quinolines as potential fluorescent probes for DNA detection: Microwave-assisted synthesis, spectroscopic characterization and crystal structure determination. Dyes and Pigments. 2011;91;79-88. DOI: :10.1016/j.dyepig.2011.02.003.

[41] Kai J, Si-Hong C, Shi-He L, Chu-Ming P, Xin-Yan W, Zhao-Yang W. Concise design and synthesis of water-soluble fluorescence sensor for sequential detection of Zn(II) and picric acid via cascade mechanism. Dyes and Pigments. 2019;167:164-173. DOI: 10.1016/j.dyepig.2019.04.023.

[42] Lanying W, Xiaogang Z, Fengmei L, Zuxun Z. Microwave-Assisted Solvent-Free Synthesis of Some Styryl Dyes with Benzimidazole Nucleus. Synthetic Communications. 2004;34:2245-2252. DOI: 10.1081/SCC-120038508.

[43] Sambathkumar S, Priyadharshini S, Fleisch M, Bahnemann DW, Gnana Kumar G, Senthilarasu S, Renganathan R. Design and synthesis of imidazole-triphenylamine based organic materials for dye-sensitized solar cells. Materials Letters. 2019;242:28-31. DOI: 10.1016/j.matlet.2019.01.091.

[44] Chinna B, Ravindra MK, Ashok Kumar K, Hemraj Y, Sivalingam R, Surendra S, Nabeen KS, Mahadevan KM, Veeranjaneya R, Arifullah M. Microwave-assisted synthesis of imidazolyl fluorescent dyes as antimicrobial agents. Journal of Materials Research and Technology. 2020;9:6900-6908, DOI: 10.1016/j.jmrt.2020.01.011.

[45] Yi-Feng S, Wei H, Chang-Gui L, Yi-Ping C. The synthesis, two-photon absorption and blue upconversion fluorescence of novel, nitrogen-containing heterocyclic chromophores. Dyes and Pigments. 2009;81:10-17. DOI:10.1016/j.dyepig.2008.08.003.

[46] Manuela M, Massimo Z, Laura F, Massimo G, Alberto Z, Magda M, Raffaella C, Stefano T, Stefano T, Michele M. Thienopyrrolyl dione end-capped oligothiophene ambipolar semiconductors for thin film- and light emitting transistors. Chemical Communications. 2011;47:11840-11842. DOI: 10.1039/c1cc14179a.

[47] Massimo Z, Francesca DM, Antonella C, Giuseppe G, Manuel P,

Fabio DS, Giovanna B. Microwave-Assisted Synthesis of Thiophene Fluorophores, Labeling and Multi-labeling of Monoclonal Antibodies, and Long Lasting Staining of Fixed Cells. Journal of American Chemical Society. 2009;131:10892-10900. DOI: 10.1021/ja902416s.

[48] Arthur DH, Casper MM, Gregory CW. Synthesis of an H-aggregated thiophene-phthalimide based small molecule via microwave-assisted direct arylation coupling reactions. Dyes and Pigments. 2014;102:204-209. DOI: 10.1016/j.dyepig.2013.10.046.

[49] Issah Y, Nurgul S, Zeynel S. Improved one-pot synthetic conditions for synthesis of functionalized fluorescent coumarin-thiophene hybrids: Syntheses, DFT studies, photophysical and thermal properties. Tetrahedron 2019;75:2143-2154. DOI: 10.1016/j.tet.2019.02.034.

[50] (a) Lung-chang L, Andrew TH. Synthesis of soluble functional dye phthalocyanines and perylene tetracarboxylic derivatives by microwave irradiation and their photoelectric performances. Journal of Porphyrins Phthalocyanines. 2003;7:565-571. DOI: 10.1142/S1088424603000719; (b) Mark OL, Andrew TH. Microwave-assisted synthesis of phthalocyanine–porphyrin complex and its photoelectric conversion properties. Journal of Organometallic Chemistry. 2004;689:2450-2455. DOI: 10.1016/j.jorganchem.2004.05.008.

[51] (a) Shaabani A, Maleki-Moghaddam R, Maleki A, RezayanAH.Microwave-assistedsynthesis of metal-free phthalocyanine and metallophthalocyanines. Dyes and Pigments. 2007;74:279-282. DOI: 10.1016/j.dyepig.2006.02.005; (b) Galina AS, Evgeny VA, Vladimir GV, Evgeny AL, Evgeny VT, Vitalij DS. Microwave-Assisted Synthesis of Phthalocyanine Zinc Complexes Derived from Aminotricyanobiphenyl Based Azo Dyes, Macroheterocycles. 2016;9:80-88. DOI: 10.6060/mhc151192s.

[52] (a) Brown GH. Photochromism techniques of chemistry vol-3. John Wiley & Sons, New York. 19719; (b) Ilyas W, Celeste B, Nicholas JT, Koji N. Porphyrins as photosensitizers to enhance night vision. Journal of American Chemical Society. 2004; 126:9892-9893. DOI:10.1021/ja0486317; (c)Henri B, Laurent L, Heinz D. Organic photochromism. (Iupac technical report) Pure Applied Chemistry. 2001; 73:639-665. DOI:10.1351/pac200173040639; (d) Appenroth K, Reichenbacher M, Partzold R. Thermochromism and photochromism of aryl-Substituted acyclic azines. II: Photokinetics. Journal of Photochemistry. 1980;14:39-50. DOI: org /10.1016/0047-2670 (80) 85066-0; (e) Appenroth K, Reichenbacher M, Partzold R. Thermochromism and Photochromism of aryl-substituted acyclic azines, III: Investigations on the mechanism of photochemical Isomerization. Journal of photochemistry. 1980;14:51-60. DOI: org/10.1016/0047-2670 (80) 85067-2; (f) Chu NYC. Photochromism of Spiroindolinonaphthoxazine. Photophysical Properties. Canadian Journal of Chemistry. 1983;61:300-305. Doi:org/10.1139/v83-054.

[53] Chung-Chun L, Jo-Chin W, Andrew TH. Microwave-assisted synthesis of photochromic spirooxazine dyes under solvent-free condition. Materials Letters. 2004;58:535-538. DOI: 10.1016/S0167-577X(03)00541-X.

[54] Xiaoli Y, Baojie Y, Yuanyuan L, Hongjun Z. Microwave-assisted synthesis of novel spirooxazines and their photochromic behaviors in polymer matrices. Optoelectronics And Advanced Materials-Rapid Communications 2012;6:1146-1152.

[55] Wei-Woon WL, Leong-Ming G, Teck-Peng L. Microwave-Assisted synthesis of photochromic fulgimides. Journal of Photochemistry and Photobiology. 2007;185:106-109. DOI: 10.1016 /j.jphotochem.2006.05.018.

[56] Xiaoliu L, Cuiying L, Shichong P, Hua C, Pingzhu Z. Convenient. Microwave-assisted, one-pot synthesis of photochromic fulgimides bearing reactive groups. Synthetic Communications. 2010;1:157-166: DOI: 10.1080/00397910902964809.

[57] Sivasankaran N, Chinnusamy S, Sengodan S, Palaninathan K . Microwave-assisted synthesis of photochromic fulgides. Journal of Chemical Science. 2010;122:183-188.

[58] Nosrat OM, Somayyeh R, Meysam PN. Microwave-assisted synthesis and photochromic properties of new azo-imidazoles. Dyes and Pigments. 2017. DOI: 10.1016/j.dyepig.2017.04.053.

[59] Alexis P, Kane D, Lara W. Synthesis of stereochemically-biassed spiropyrans by microwave-promoted, one-pot alkylation-condensation. Organic and Biomolecular Chemistry. 2018;16:7245-7254. DOI: 10.1039 /C8OB01996G.

[60] Verma J, Islam, SM, Verma A, Protaseko V, Jena D. Nitride LEDs based on quantum wells and quantum dots. Woodhead publishing series in electronic and optical materials. 2018: 377-413. DOI: 10.1016/B978-0-08-101942-9.00011-3.

[61] Mao-Kuo W, Chii-Wann L, Chih-Chung Y, Yean-Woei K, Jiun-Haw L, Hoang-Yan, L. Emission characteristics of organic-light-emitting diodes and organic thin-films with planar and corrugated structures. International Journal of molecular Sciences. 2010;11:1527-1545. DOI: 10.3390/ijms11041527.

[62] John B, Frederic C, Yasushi Y, Jean PB. Photochromism of an Indolylfulgide trapped in hybrid sol-gel matrix. Chemistry Letters. 1998;27: 359-360. DOI: 10.1246/cl.1998.359.

[63] Yuichi K, Atsushi N, Kazuo T, Yoshiki C. Efficient simultaneous emission from RGB-emitting organoboron dyes incorporated into organic-inorganic hybrids and preparation of white light-emitting materials. Journal of Materials Chemistry C. 2013;1:4437-4444. DOI: 10.1039/c3tc30276h.

[64] Jefri, Deana W. Microwave-assisted transformation of *N, N*-Diphenylamine as Precursors of Organic-light Emitting Diodes (O-LED). American Institute of Physics. 2015; 1677. DOI: org/10.1063/1.4930716.

[65] Dumitru P, Andrei D, Aurelian R, Luminita M. Microwave-assisted synthesis of an alternant poly(fluorine-oxadiazole). Synthesis, properties and white light-emitting devices. Polymers. 2019;11:1562-1579.DOI: org/10.3390/polym11101562.

Chapter 4

Doping of Semiconductors at Nanoscale with Microwave Heating (Overview)

Sandhya K. M., Litty Thomas Manamel and Bikas C. Das

Abstract

Incorporation of dopants efficiently in semiconductors at the nanoscale is an open challenge and is also essential to tune the conductivity. Typically, heating is a necessary step during nanomaterials' solution growth either as pristine or doped products. Usually, conventional heating induces the diffusion of dopant atoms into host nanocrystals towards the surface at the time of doped sample growth. However, the dielectric heating by microwave irradiation minimizes this dopant diffusion problem and accelerates precursors' reaction, which certainly improves the doping yield and reduces processing costs. The microwave radiation provides rapid and homogeneous volumetric heating due to its high penetration depth, which is crucial for the uniform distribution of dopants inside nanometer-scale semiconducting materials. This chapter discusses the effective uses of microwave heating for high-quality nanomaterials synthesis in a solution where doping is necessary to tune the electronic and optoelectronic properties for various applications.

Keywords: conductivity tuning, microwave heating, doping at the nanoscale, dopant diffusion, nanocrystals, solution growth

1. Introduction

The discovery of microwave cooking by Percy Spencer marked the dawn of a new era in microwave heating technology, which has gained huge attention in the scientific areas, especially in synthetic chemistry [1]. Numerous factors enabled the microwave technique to become a breakthrough technology in the complex synthesis process [2]. The significance of microwave heating for the synthesis of high-quality semiconducting nanomaterials, pristine or doped ones, is a subject that needs to be profoundly studied and explored due to its capability to revolutionize the semiconductor industry. The synthesis of high-quality nanocrystals primarily relies on controlled reactions of molecular precursors in a liquid medium at an adequate temperature in the presence of stabilizing agents [3, 4]. Most of the synthetic methods such as wet chemical process [5], emulsion methods, anti-solvent precipitation methods [6], have studied only the impact of the chemical process and parameters, on the properties of as-synthesized nanocrystals. Of late, the effect of additional external stimulations like microwave irradiation [7, 8], ultraviolet/visible light irradiation, ultrasound, etc. is also studied [5]. Microwave heating increases the rate of reaction, thereby considerably decreasing the reaction time without altering the kinetics and chemical reaction [9–11]. The rate accelerations caused by "specific microwave effect" as well as "non-thermal effects" have to be considered in the microwave heating mechanism. Baghbanzadeh *et al*. propose that microwave dielectric heating can be termed as "specific microwave effects" by which one can achieve rate accelerations that cannot be attained by the conventional methods [12]. In the case of "non-thermal microwave effects", the heating mechanism arises as a result of the direct interaction of microwaves with specific molecules or materials in the reaction medium [2, 12]. Jacob *et al.* report that the enhancement rate of reaction with microwave heating compared to conventional heating is mainly due to the thermal effects which arise due to three significant factors. Firstly, the localized heating effect is a consequence of superheating phenomena due to the abundant ions present in the medium. Secondly, the molecular agitation due to lag of dipole, in following the fast-moving EM wave. Thirdly, increase of diffusion rate of reactants [13].

Earlier, the synthesis of high-quality semiconducting quantum dots was very tough and the process of doping at this length scale makes it even more challenging. Erwin *et al.* reported that the difficulty in doping at nanoscale regime is due to the difference in mechanisms involved in doping at bulk and at the nanoscale, while other reports in literature claim the process of 'self-purification' as the leading cause for de-doping during the growth process [14]. The major daunting challenges arise mostly due to the lack of a comprehensive understanding of all the fundamental mechanisms associated with dopants incorporation and the absence of reliable synthetic procedures where the temperature-dependent dopant impurity atoms diffusion will be minimal [15]. Another challenge involved in doping at the nanoscale is the inherent statistical inhomogeneity of dopants among the nanocrystals. The doped nanomaterials always tend to exhibit a broad range of dopant populations per nanocrystal, which results in effective inhomogeneity in concentration of dopants among nanocrystals. Providing a uniform and instantaneous heating during the reaction process can minimize this problem to a great extent [16]. In this context, microwave heating became the suitable thermal energy source for doping the semiconducting nanocrystals, as it provides rapid and instantaneous heating. Short reaction time, faster reaction rate, uniform volumetric heating, cost-effective and eco-friendly method are the other remarkable features which make microwave heating a prime superior choice over other conventional methods of heating like a hot plate, oil bath, etc. [8]. Moreover, heating by means of conventional methods always results in 'a self-purification' mechanism where the dopants are diffused towards the surface of nanocrystals at the time of growth [14]. By adopting microwave-assisted techniques the aforementioned problems encountered while doping at the nanoscale can be eliminated to a great extent and the synthesized products are found to excel both in quality as well as quantity [17].

2. Physics behind the microwave heating

Microwave is an electromagnetic wave having low energy with a wavelength and a frequency in the range of 1 m to 1 mm and 300 MHz to 300 GHz, respectively as shown in **Figure 1**. Mainly, the laboratory and household microwave oven operate at a frequency of 2.45 GHz, which corresponds to a wavelength of 11.2 cm. It can travel at the speed of light (~30 cm/nanosecond) like any other electromagnetic wave and consists of electric and magnetic fields oscillating in a direction perpendicular to each other. One can also define it as a Multiphysics phenomenon in which the heating arises due to interaction between matter and electromagnetic radiation.

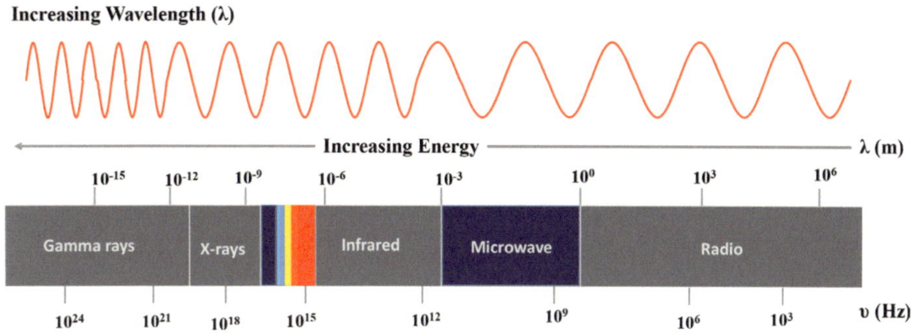

Figure 1.
Schematic representation of the electromagnetic spectrum in terms of wavelengths and frequencies.

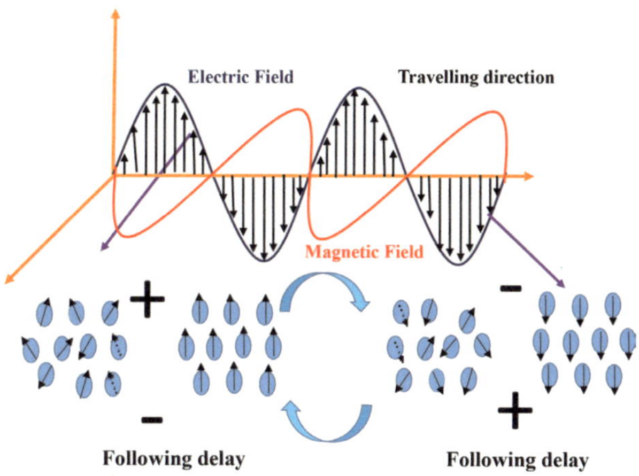

Figure 2.
Schematic diagram of the interaction of an electric component of the microwave radiation with matter.

In contrast to other conventional methods of heating, here the medium itself gets self-heated as a result of the alignment of molecular dipoles present in it with respect to the field associated. The electric and magnetic components in microwave interact with matter in different manners as discussed below [1].

2.1 Influence of electric field component

The polar molecules are sensitive to an electric field, and thus as a result of force exerted by the field on the charged particles, they start to migrate or rotate in order to align along the field (**Figure 2**). Since the electric and magnetic components reverse direction rapidly with a frequency of 2.45 GHz, the electric dipoles have no time to orientate to the direction of electric filed. As a result, there occurs the angle between the orientation of the dipoles in space and the direction of the electric field and the energy loss by the dipoles occurs resulting in to rise of dielectric heating. Reflection, absorption, and transmission are the three modes by which the medium reacts to the electromagnetic waves, either in a single or combined fashion [18]. The effective dielectric loss factor for the dielectric heating can be expressed in terms of dipolar polarization, ionic conduction as follows

$$\epsilon''_{ff} = \varepsilon''_{polarisation} + \epsilon''_{dipolar} = \epsilon''_{interfacial} + \frac{\sigma}{\omega\epsilon_o} + \epsilon''_{dipolar} \quad (1)$$

where $\epsilon''_{polarisation}$, $\epsilon''_{dipolar}$, $\epsilon''_{interfacial}$, σ, ω, and ϵ_0 represent the polarization dielectric loss, dipolar dielectric loss, interfacial dielectric loss, electrical conductivity (S/m), angular frequency (Hz), and permittivity of free space (8.854×10^{-12} F/m), respectively [19].

2.2 Influence of magnetic field component

Like an electric field, a magnetic field interacts too with matter and induces heat through magnetic loss, joule heating, and so on. However, sufficient studies apart from dielectric heating are still very rare. Meanwhile, Cheng *et al.* reported that magnetic loss contributes significantly to microwave heating compared to dielectric heating [20]. The necessary physical processes generating heat energy as a result of interaction between material medium and the magnetic field component are the eddy current loss, hysteresis loss, and magnetic resonance loss [21, 22]. The overall losses that constitute the effective magnetic permeability ($\mu''_{effective}$) we may define as

$$\mu''_{effective} = \mu''_{hysteresis} + \mu''_{eddy\ current} + \mu''_{residual} \quad (2)$$

where $\mu''_{hysteresis}$, $\mu''_{eddy\ current}$, and $\mu''_{residual}$ represent the hysteresis magnetic loss (H/m), eddy current magnetic loss (H/m), and residual magnetic loss (H/m), respectively [18].

3. Doping at the nanoscale and microwave heating

Microwave heating has been the subject of interest for doping semiconductors at nanoscale owing to its ability to control the synthesis process explicitly. Apart from being cost-effective, the dielectric heating by microwave irradiation minimizes the dopant diffusion problem and provides quick reaction among precursors. The process of nucleation and growth of nanocrystals have been described in theories like LaMer burst nucleation [23], Watzky and Finke's slow nucleation followed by autocatalytic growth [24], and LSW theory, etc. [25, 26]. Nucleation is the process where nuclei act as a template for nanocrystal growth. Uniform formation of nuclei throughout the growth medium defined as 'homogeneous nucleation' can be easily and efficiently achieved by microwave irradiation in contrast to conventional methods of heating. Volumetric heating provided by microwave irradiation raises

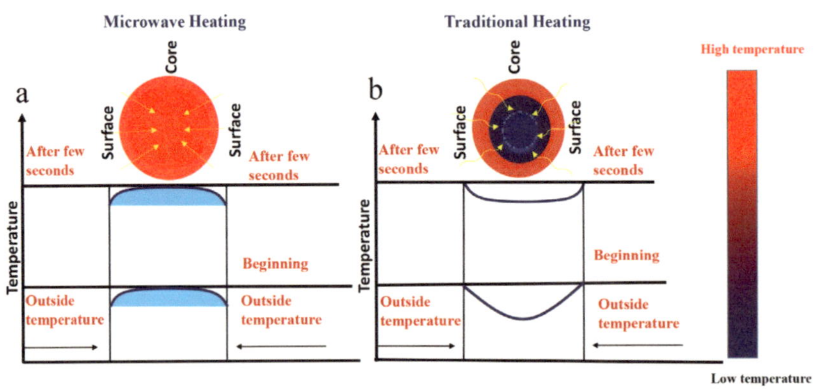

Figure 3.
Schematic illustration of main differences between the microwave heating (a) and traditional heating method (b).

the internal temperature of the whole medium simultaneously and homogeneously as illustrated in **Figure 3**. This favors a quick nucleation process which results in solution supersaturation leading to homogeneous nucleation. Microwave-assisted technique aids to measure, manipulate, and thereby optimize the nucleation process and parameters that in turn influences the stability of the synthesized particles along with an added advantage of automatic data recording [12]. Efficient doping is determined by the surface morphology and shape of nanocrystals and the presence of surfactants in the reaction medium. Temperature plays a significant role in molding the aforementioned factors [27]. This demands the necessity for a proper thermal energy source like microwave heating while synthesizing nanocrystals. High penetration depth (*d*) offered by microwave heating is yet another factor that distinguishes it from the conventional methods of heating. It is defined as the distance at which the microwave power reduces to 1/*e* of its incident power. It has inverse proportionality with oscillating frequency, dielectric, and magnetic loss factor. The formula for determinate a penetration depth (*d*) may be written as

$$d = \frac{1}{\alpha} = \sqrt{\frac{2}{\omega^2 \mu_0 \mu' \varepsilon''_{eff} \varepsilon_0}} \quad (3)$$

where α is the absorption coefficient of microwaves, ω is the oscillating frequency, μ_0 is the permeability, ε''_{eff} is the dielectric loss factor, ε_0 is the vacuum permittivity, μ' is the magnetic loss factor, etc. [19]. In addition to the above-mentioned factors, the specific interaction of electromagnetic wave with the precursors plays a significant role in the formation of nanocrystals.

The efficient absorption of the EM wave by the solvent is determined by its loss tangent factor. It is defined as the ability of a material to convert electromagnetic energy into heat energy at a given frequency and temperature [1]. A high value is desired for maximum absorption, however, heating aided by microwave radiation is achievable even in the presence of a low tan (δ) solvent provided there exists either a polar reactant or reagent such that the overall dielectric nature of the reaction medium favors the microwave heating. In the case of conventional heating methods, the transfer of heat is slow and inefficient, resulting in a huge temperature gradient owing to the different thermal conductivity of materials. However, in the case of microwave radiation, there is a direct coupling between the microwave energy and the molecules resulting in core volumetric heating. The most commonly used frequency of the microwave is 2.45 GHz, possessing an energy of 0.0016 eV, which is lower than that of Brownian motion and therefore insufficient to break the bonds. This property of microwaves makes them incapable of carrying out any unwanted reactions and thereby solely ensuring effective doping at nanoscale materials.

3.1 Doping of semiconducting nanomaterials

Nanocrystals are broadly classified as nanoparticles and quantum dots. Generally, tiny particles of a dimension of 100 nm or below are termed as nanoparticles. However, quantum dots (QDs) are a class of nanomaterials with their charge carriers confined in all three dimensions of the length scale of exciton Bohr radius [28]. While doping the QDs, the dopants have a high tendency to come out of it due to the thermal diffusion because their size is in the nanometer range. This problem can be resolved greatly by having a comprehensive idea about the various mechanisms involved during doping and following a proper synthesis process [29]. However, various properties, including optical, magnetic, and electronic, of semiconducting quantum dots can be tailored in a desired fashion by the incorporation of impurity dopant atoms [30]. Moreover, this can also generate some new physical properties, including spin-polarizable excitonic photoluminescence, exciton storage, excitonic magnetic polaron formation, and magnetic circular dichroism so on. The proper incorporation of impurity atoms into the semiconducting QDs is a tough job but can be identified by observing the following features like red-shifted PL emission and large Zeeman splitting of excitonic excited states that are a result of strong exchange coupling between dopant and the carrier [31, 32]. In the year 2000, Mikulec *et al.* reported the most significant result on QDs doping; in which they reported manganese (Mn) doped CdSe nanocrystals with the evidential result obtained from electron paramagnetic resonance (EPR) [33]. Later, a variety of doped semiconducting material were reported by tailoring both the host atoms such as ZnS, PbS, MgO, Al_2O_3, α-Fe_2O_3, CdS, ZnSe, etc. and dopant atoms such as Mn, Cu, Ag, Fe, Zn, Cr, Er, etc. [34, 35]. However, there is a limitation to select the host system and the respective dopant atoms. Suppose, incorporation of Mn into nanocrystals of CdS and ZnSe easy but not into CdSe even though the bulk solubility almost equal to 50% for all three [27, 36].

Depending on dopants' diffusivity, the dopant precursors are injected at different time intervals, suppose along with the host precursors or at the time of nucleation or growth as shown in **Figure 4** [37]. The major problem involved in doping at the nanoscale is that many dopants fail to be incorporated within the host lattice and instead get adsorbed on the surface [38]. High formation energy for defects renders the impurity atoms to be thermodynamically unstable, resulting in the expulsion of dopants from the host lattice, in turn leading to self-purification [14, 27, 39]. Apart from thermodynamics, kinetics also play a significant role in determining the stability of added impurities in solution phase synthesis. Chen *et al.* have reported a detailed study regarding all the elemental processes involved with doping, such as surface adsorption, lattice incorporation, lattice diffusion, and lattice ejection as represented schematically in **Figure 5** [40]. Maintenance of appropriate temperature is a crucial factor even in the phenomena mentioned above.

The high cost of commercially doped QDs is one reason that limits its wide range of applications. Therefore, cost-effective synthesis protocols need to be developed to produce high-quality doped QDs. This limitation and the ones mentioned above are lifted off using microwave heating for doping the QDs. It is also found to be an economical and eco-friendly method in line with green chemistry. Now let us discuss some semiconducting QDs systems where doping has been performed with microwave heating technique.

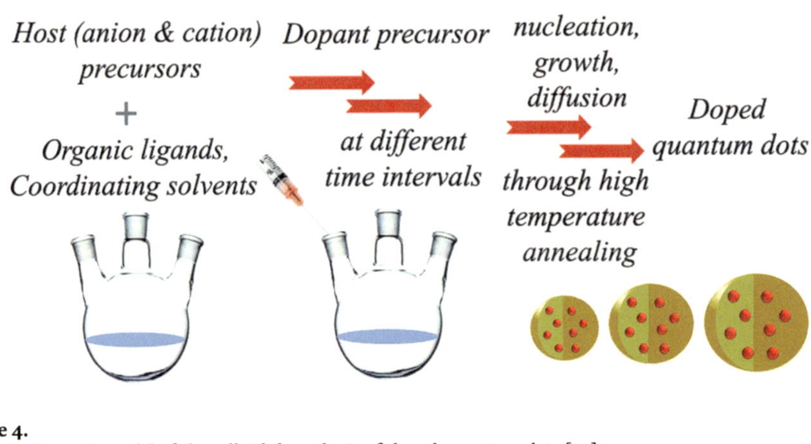

Figure 4.
General schematic model of the colloidal synthesis of doped quantum dots [37].

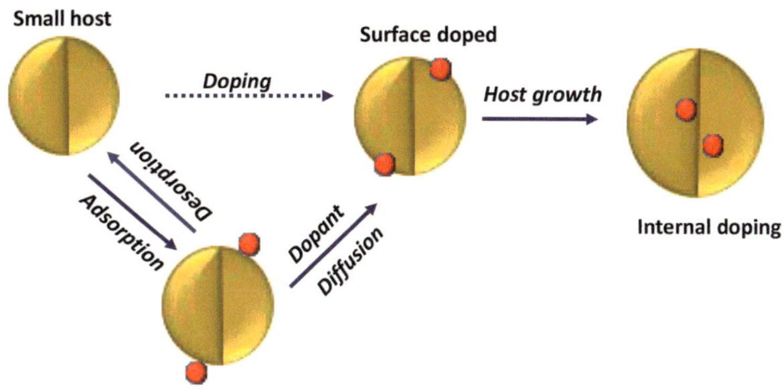

Figure 5.
Schematic diagram showing temperature-dependent dopant lattice diffusion [37].

3.1.1 Mn-doped CdSe quantum dots

CdSe QDs is n-type intrinsically, and a flagship candidate in nanoscale research history shows several novel properties as a member of the II-VI binary semiconductor group. It was attractive to the researchers to demonstrate various optoelectronic applications as its energy band overlaps nicely with the solar energy spectrum [41]. The fundamental properties of CdSe are enhanced via doping, which further increases its demand in the semiconductor industry. However, doping of CdSe by Mn^{2+} ions is challenging due to the self-purification effect, as reported by Erwin *et al.* [27, 29, 42]. The doping process is mainly governed by the surface kinetic effect. Microwave heating helps one to have exquisite control over this surface kinetics that eases the doping process.

Meladom *et al.* developed a robust synthesis protocol for efficient doping of Mn^{2+} into CdSe QDs in an aqueous medium with mild microwave heating as a final step [17]. A household microwave oven was used to heat the CdSe QDs solution for 60 seconds duration with the set point of 450 W (operational frequency 2.45 GHz). This heating step was repeated three times by giving 5 minutes intervals. The motivation was to tune the electrical conductivity of CdSe QDs thin film by varying doping concentration only as the size of QDs kept similar for all the samples. Microwave heating improves the quality of QDs in terms of optical properties, which was confirmed by recording UV–vis absorbance and photoluminescence both excitation and emission spectra, as shown in **Figure 6(a)** and **(b)**, respectively. In all the cases, peak intensities were enhanced and bandwidth reduced, which indicates the reduction of surface defects of QDs. The chemical composition of the doped CdSe QDs sample was confirmed with X-ray photoelectron spectroscopy (XPS), energy-dispersive X-ray spectroscopy (EDS), and inductively coupled plasma - atomic emission spectroscopy (ICP-AES) measurements data. XPS result confirmed the efficient incorporation of Mn atoms as dopants inside the host CdSe QDs (**Figure 7**).

3.1.2 More examples on doped binary nanocrystals

Microwave-assisted synthesis has also been utilized by many research groups around the world to dope various other binary II-VI semiconductor-based nanocrystals. Molaei *et al.* reported the synthesis of copper (Cu) doped ZnSe nanocrystals in the aqueous medium to study the doping effect on the optical properties [43]. Synthesis of Mn^{2+} ion-doped ZnS quantum dots was reported by Joicy *et al.* using a rapid microwave irradiation step without any surfactants, which showed photocatalytic activity by observing photodegradation of methyl orange dye under

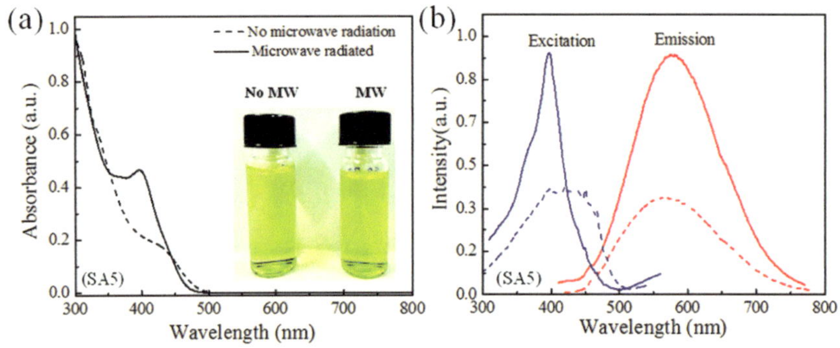

Figure 6.
Electronic UV–vis (a) and photoluminescence (b) spectrum of Mn^{2+} doped CdSe QDs sample with and without microwave irradiation [17].

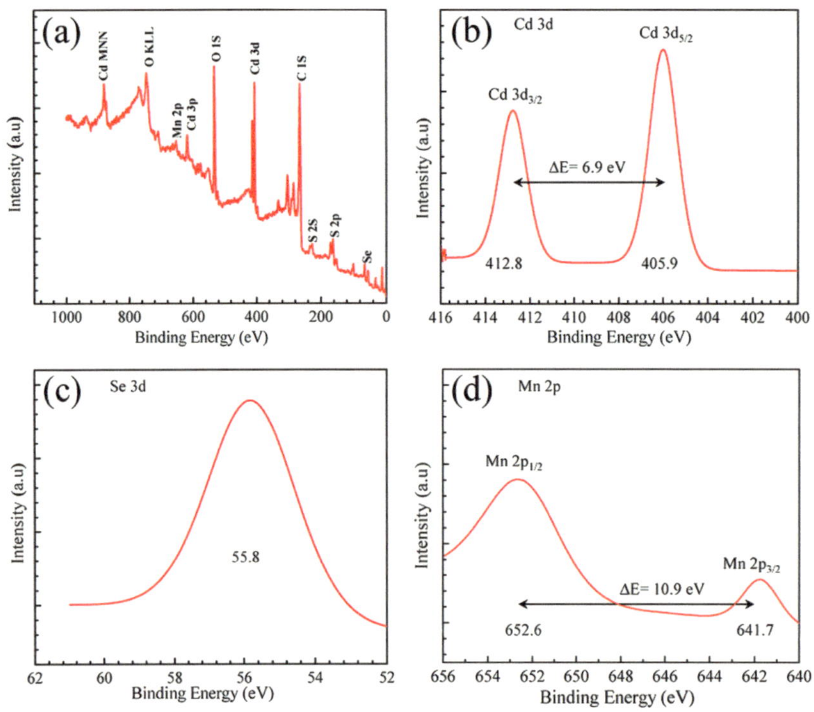

Figure 7.
(a) Survey scan of the X-ray photoelectron spectrum of 2% Mn^{2+}-doped CdSe QDs. (b, c) high-resolution spectrum of Cd 3d electrons depicting doublet splitting with binding energies separated by 6.9 eV and Se 3d, respectively. (d) the high-resolution spectrum of Mn 2p core electrons showing doublet splitting with binding energy separated by 10.9 eV [17].

UV light irradiation [44]. Here, the zinc blende crystal phase of ZnS was important for the efficient incorporation of Mn atoms. In the same year, Zhu *et al* reported the synthesis of of Mn-doped ZnS via green and rapid microwave-assisted approach and they also developed indapamide drug detector by recording room-temperature phosphorescence (RTP) with that doped material [45].

Later, Zhang *et al.* reported the aqueous synthesis of Mn and Cu doped ZnSe QDs by microwave radiation with higher quantum yields (QYs) and they have further extended this work to grow the white-light-emitting ZnSe/ZnS core/shell QDs via the co-doping of Mn and Cu [46]. Lead sulphide (PbS) QD are still emerging various applications in optoelectronics and its property was further tuned with a silver (Ag) atom doping. It is also reported by Shkir *et al.* that the bandgap of PbS QDs was increased with Ag atom incorporation, which was predicted without mentioning the influence of size variation between the samples used [47]. Recently, another work reported on facile microwave synthesis of CdS quantum dots doped with Cr atoms as impurity doping and they have studied various properties like structural, opto-dielectric, electrical, and so on [48].

3.1.3 Doped oxide nanomaterials

Various metal oxides nanomaterials play a major role in the development of different novel daily life applications in the fields of display, sensors, medicine, biomedical devices, agriculture, information technology, optical, energy, electronics, and so on. Efforts are ongoing to tune their properties and applications further with incorporating impurity as dopants. For that reason, microwave heating based synthesis protocols is developing as a potential alternative to the conventional heating based growth process. Kar *et al.* developed a microwave synthesis of rare-earth element Eu^{3+} doped tin oxide (SnO_2) to tune the optical and electrical properties of the host [49]. Jamatia *et al.* reported the microwave-assisted synthesis of Fe doped ZnO nanoparticles to show their application in polymer light-emitting diodes [50]. The wurtzite hexagonal crystal phase of ZnO nanoparticles and incorporation of the Fe dopant into the host ZnO crystal lattice was confirmed via X-ray diffraction analysis. This report claimed that the bandgap modification of ZnO via Fe doping is estimated from the Tauc plot without considering the influence of size. Similarly, many spinel structured metal oxides were also doped with different transition metal ions via microwave heating based synthesis technique with tunable structural, morphological, optical, vibrational, and magnetic properties and different potential applications like phosphor-based forensic testing and many more [51–54]. Interestingly, Er^{3+} doped α-Fe_2O_3 and Fe doped TiO_2 nanoparticles were synthesized successfully with the help of microwave heating to study their crystal structure and optical properties [55, 56]. Recently, Yathisha *et al.* reported Zn^{2+} doped MgO nanoparticles utilizing microwave combustion route to study the influence on photovoltaic properties [57]. Therefore, microwave heating could explore further as a low-cost alternative synthesis protocol to design a new variant of nanomaterials.

3.1.4 Lanthanide ion doped lanthanum trifluoride (LaF3)

Lanthanum trifluoride (LaF_3) is an ionic compound that is utilized as core-shell-up conversion nanoparticles (UCNPs) for different filed of applications like sensing, biomedical, and solar cells. Tek *et al.* reported Yb^{3+} ion-doped (active) and undoped (inert) LaF_3 shell coatings on a 20% Yb, 2% Tm codoped hexagonal phase LaF_3 core with the help of microwave -assisted synthesis route [58]. They observed higher optical enhancement of inert shell compared with the active shell at all prominent emission peaks, which is explained with the energy band diagram indicating the energy transfer pathways for the Yb^{3+} and Tm^{3+} – co-doping (**Figure 8**).

3.2 Doping of carbon based nanomaterials

Carbon-based materials like graphene can also be doped via microwave (MW) heating. Since the nanocarbon materials are found to be sensitive to microwave radiation [59, 60], the technique of MW heating was employed in the modification of graphene materials. It is also reported that by the use of microwave heating, hollow carbon nanospheres can be synthesized within a short time which can be effectively used as a host material for doping [61, 62]. **Figure 9** shows the

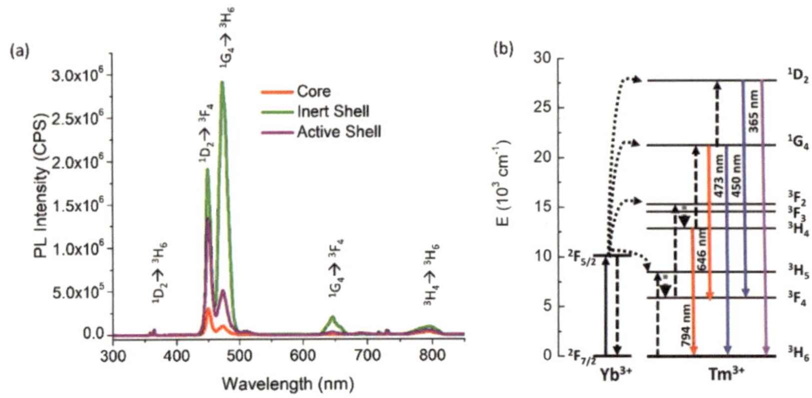

Figure 8.
(a) UCPL data for core, active shell, and inert shell nanoparticles under 980 nm continuous-wave excitation. (b) Energy diagram showing the corresponding transition of UCPL of (a) where the energy transfer pathways for the Yb^{3+} and Tm^{3+}-codoped up conversion nanoparticles are depicted [58].

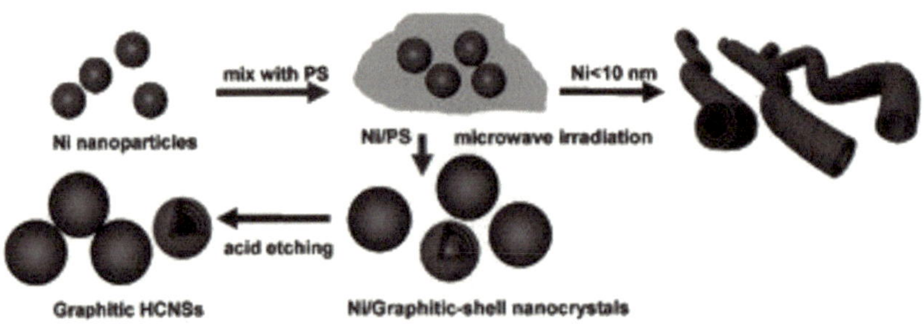

Figure 9.
Microwave-induced synthesis of Ni/graphitic-shell nanocrystals and graphitic hollow carbon Nano spheres [61].

microwave-assisted approach to prepare metal/graphitic shell nanocrystals and CNT in a very short time using ordinary carbon precursor.

The microwave-assisted technique facilitates the growth of heteroatom-doped graphene with better catalytic activity as well. Nitrogen doping up to 8.1% on graphene was achieved by Kwang *et al* within a minute with the aid of the microwave radiation. The binding configuration of nitrogen over graphitic basal planes can be varied with the irradiation power of microwave. The conductivity enhancement upto 300 Scm^{-1} was obtained in this case in comparison to nitrogen-doped via arc discharge method, nitrogen plasma process, etc. showing a lesser conductivity [63]. The dielectric heating of MW induces a high energy state that helps the graphitic basal plane to accommodate the dopants in order to convert graphite to N-doped graphene. The selective dielectric heating, which arises due to the difference in the dielectric constants of solvent and reactant can enhance the efficiency of doping without the rise of a thermal gradient [64]. The solid phase microwave-assisted synthetic method is adopted for the large-scale production of N-doped carbon nanodots (CNDs) using different citric acid/urea (C/U) weight ratios, which result in size variation of CNDs as shown in **Figure 10** with the transmission electron microscope (TEM) images. The dopant ion concentration can be varied in a precise manner that results in N-doped graphene QDs and graphitic-carbon nitride quantum dots (g-CNQD). The doped material is found to exhibit a 38.7% quantum yield due to the presence of N and O rich edge groups resulting from the interaction of microwave on graphene [65].

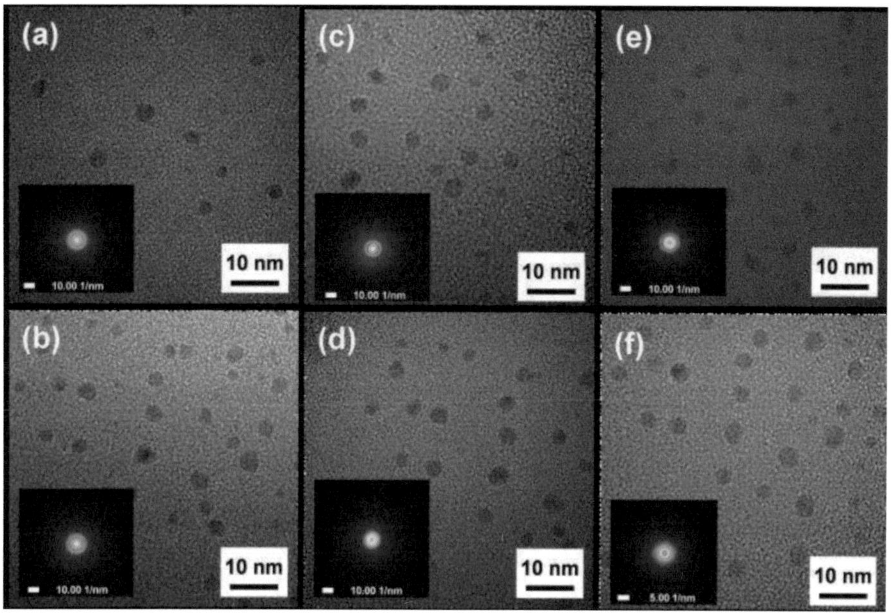

Figure 10.
HR-TEM micrographs of N-doped carbon nanodots (CNDs) samples prepared using the SPMA method for different citric acid /urea (C/U) weight ratios of 3/1 (a), 2/1 (b), 1/1 (c), 1/1.5 (d), 1/2 (e), and 1/3 (f). Inset shows the corresponding selected-area diffraction pattern [65].

4. Electrical and memristor property of Mn^{2+}doped CdSe QDs

Huge enhancement in the conductivity of microwave-assisted doped QDs has been reported in many pieces of literature. Microwave heating enables the tuning of electrical conductivity in a desired manner by proper incorporation of dopants into the desired locations of the host material. This is evidenced by the rise in the electrical conductivity to the order of 10^4 for 2% Mn^{2+} doped CdSe over undoped one as shown in **Figure 11(a)** [17]. Here, the conduction mechanism is controlled by the electric field-assisted thermal ionization of trapped charge carriers in CdSe QDs as described in Poole–Frenkel effect as shown in **Figure 11(b)** [66]. The bandgap has no role in the conductivity and the observed colossal conductivity enhancement is solely due to the concentration of Mn^{2+} dopant ions.

The STM study performed on a monolayer device of Mn^{2+} doped CdSe QDs synthesized via microwave method founds to exhibit excellent memory characteristics as described in **Figure 12** [17]. This memristor property is evident from **Figure 12(b)** where the doped CdSe QDs switched to a high conducting state at the bias of 2.5 V. It is also observed that the device switched back to its low conducting state when the tip swept towards 3 V and the ON/OFF ratio obtained was higher than 10^2. The reproducible nature of the resistive switching property over many cycles further confirms the reliability of the measurement. The threshold voltage at which the device switches to a high conducting state is found to be decreasing with an increase in the dopant concentration. Thus the notable electric bistability and the low threshold voltage of as synthesized doped CdSe QDs with the aid of simple and domestic microwave method promises its application in vivid area of future technologies which ensures minimum energy consumption per byte of the resistive data storage devices in future.

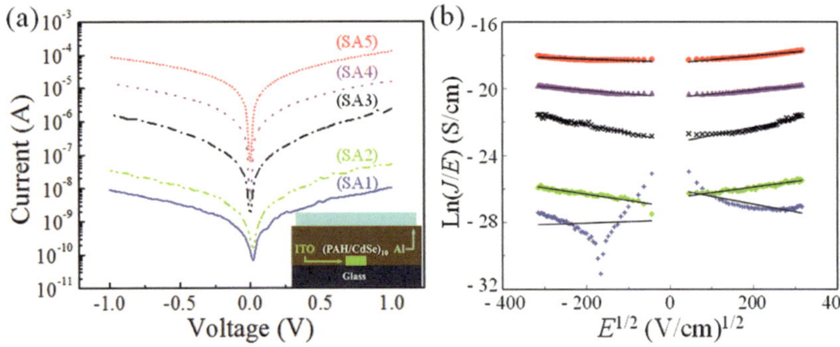

Figure 11.
(a) Current–voltage characteristics of Mn-doped CdSe QDs for the samples with varying dopant concentrations. (b) Poole–Frenkel fitting for all samples with respective straight trend line [17].

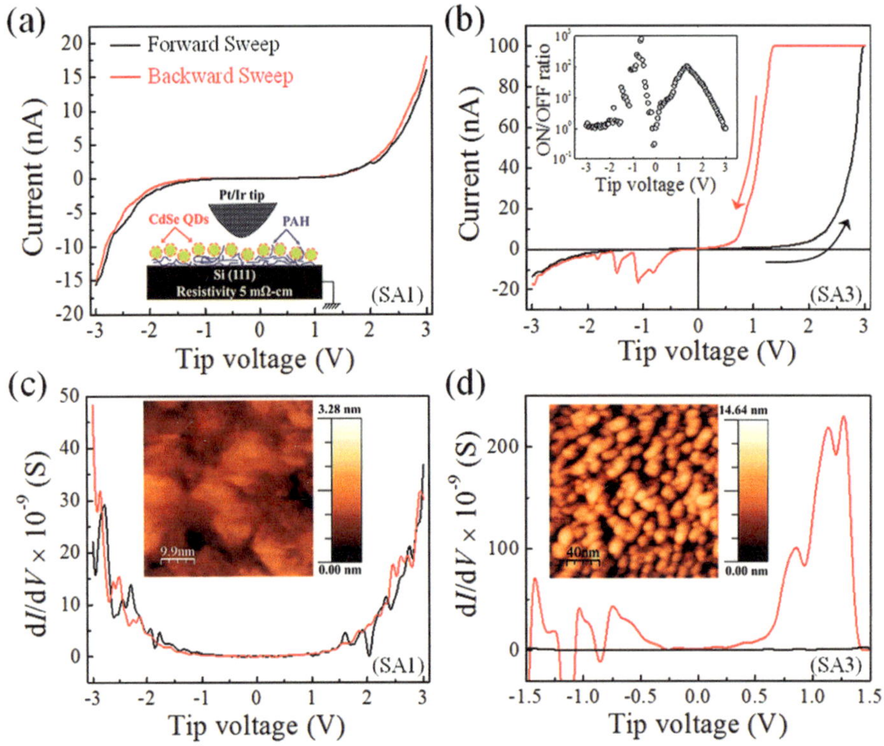

Figure 12.
(a, b) The Tunneling current–voltage (I–V) characteristics of a monolayer of undoped and 0.2% Mn^{2+}-doped CdSe QDs. Doped CdSe is showing low conducting state (OFF state, black line) and high conducting state (ON state, red line) for forward and backward voltage sweep direction respectively. (c, d) The differential conductance–voltage characteristics of a monolayer of undoped and 0.2% Mn^{2+}-doped CdSe QDs respectively in their forward (black line) and backward (red line) sweep direction. The topographic images of bare Si(111). and monolayer of undoped CdSe QDs deposited on Si(111) are shown on the insets within (c) and (d) [17].

5. Conclusion

In this chapter, we mainly discussed the incorporation of impurity dopant atoms into a host semiconducting quantum dots system very efficiently using microwave heating strategy with the help of a large number of examples from the literature. It has also been observed that the zinc blend crystal phase is very efficient for the dopant incorporation than the hexagonal one. This also reflects that microwave heating can be utilized to synthesize various classes of doped zero-dimensional (0D) nanomaterials or quantum dots of many chalcogenides, oxides, carbon dots, and more with the large numbers of dopant atoms easily and more cheaply. Literature shows that the research related to two-dimensional (2D) transition metal dichalcogenides (TMDs) is booming up due to having tunable physical, electronic, and optoelectronic properties. Therefore, it would be intriguing to grow various 2D TMDs, both intrinsic and impurity-doped, via microwave heating, which will definitely reduce cost and different health and environmental hazards.

Author details

Sandhya K. M., Litty Thomas Manamel and Bikas C. Das*
School of Physics, Indian Institute of Science Education and Research Thiruvananthapuram (IISER TVM), Maruthamala PO, Vithura, Trivandrum, Kerala, India

*Address all correspondence to: bikas@iisertvm.ac.in

References

[1] S. A. Galema, "Microwave chemistry," *Chemical Society Reviews,* vol. 26, pp. 233-238, Jun 1997.

[2] Y. J. Zhu and F. Chen, "Microwave-Assisted Preparation of Inorganic Nanostructures in Liquid Phase," *Chemical Reviews,* vol. 114, pp. 6462-6555, Jun 25 2014.

[3] M. Niederberger, G. Garnweitner, J. Buha, J. Polleux, J. H. Ba, and N. Pinna, "Nonaqueous synthesis of metal oxide nanoparticles: Review and indium oxide as case study for the dependence of particle morphology on precursors and solvents," *Journal of Sol-Gel Science and Technology,* vol. 40, pp. 259-266, Dec 2006.

[4] M. Casavola, R. Buonsanti, G. Caputo, and P. D. Cozzoli, "Colloidal strategies for preparing oxide-based hybrid nanocrystals," *European Journal of Inorganic Chemistry,* pp. 837-854, Feb 2008.

[5] P. D. Cozzoli and L. Manna, "Synthetic strategies to size and shape controlled nanocrystals and nanocrystal heterostructures," *Bio-Applications of Nanoparticles,* vol. 620, pp. 1-17, 2007.

[6] E. Reverchon, "Supercritical antisolvent precipitation of micro- and nano-particles," *Journal of Supercritical Fluids,* vol. 15, pp. 1-21, May 1 1999.

[7] I. Bilecka and M. Niederberger, "Microwave Chemistry: Towards Predictable Nanoparticle Synthesis," *Chimia,* vol. 63, pp. 581-581, 2009.

[8] I. Bilecka and M. Niederberger, "Microwave chemistry for inorganic nanomaterials synthesis," *Nanoscale,* vol. 2, pp. 1358-1374, 2010.

[9] C. O. Kappe, "Controlled microwave heating in modern organic synthesis," *Angewandte Chemie-International Edition,* vol. 43, pp. 6250-6284, 2004.

[10] F. Langa, P. DelaCruz, A. DelaHoz, A. DiazOrtiz, and E. DiezBarra, "Microwave irradiation: more than just a method for accelerating reactions," *Contemporary Organic Synthesis,* vol. 4, pp. 373-386, Oct 1997.

[11] D. Obermayer, B. Gutmann, and C. O. Kappe, "Microwave Chemistry in Silicon Carbide Reaction Vials: Separating Thermal from Nonthermal Effects," *Angewandte Chemie-International Edition,* vol. 48, pp. 8321-8324, 2009.

[12] M. Baghbanzadeh, L. Carbone, P. D. Cozzoli, and C. O. Kappe, "Microwave-Assisted Synthesis of Colloidal Inorganic Nanocrystals," *Angewandte Chemie-International Edition,* vol. 50, pp. 11312-11359, 2011.

[13] A. de la Hoz, A. Diaz-Ortiz, and A. Moreno, "Microwaves in organic synthesis. Thermal and non-thermal microwave effects," *Chemical Society Reviews,* vol. 34, pp. 164-178, 2005.

[14] G. M. Dalpian and J. R. Chelikowsky, "Self-purification in semiconductor nanocrystals," *Physical Review Letters,* vol. 96, Jun 9 2006.

[15] J. B. Li, S. H. Wei, S. S. Li, and J. B. Xia, "Origin of the doping bottleneck in semiconductor quantum dots: A first-principles study," *Physical Review B,* vol. 77, Mar 2008.

[16] J. D. Bryan and D. R. Gamelin, "Doped Semiconductor Nanocrystals: Synthesis, Characterization, Physical Properties, and Applications," *Progress in Inorganic Chemistry, Vol 54,* vol. 54, pp. 47-126, 2005.

[17] S. K. Meladom, S. Arackal, A. Sreedharan, S. Sagar, and B. C. Das, "Microwave assisted robust aqueous synthesis of Mn2+−doped CdSe QDs with enhanced electronic properties,"

Rsc Advances, vol. 8, pp. 26771-26781, 2018.

[18] D. El Khaled, N. Novas, J. A. Gazquez, and F. Manzano-Agugliaro, "Microwave dielectric heating: Applications on metals processing," *Renewable & Sustainable Energy Reviews,* vol. 82, pp. 2880-2892, Feb 2018.

[19] J. Sun, W. L. Wang, and Q. Y. Yue, "Review on Microwave-Matter Interaction Fundamentals and Efficient Microwave-Associated Heating Strategies," *Materials,* vol. 9, Apr 2016.

[20] J. P. Cheng, R. Roy, and D. Agrawal, "Experimental proof of major role of magnetic field losses in microwave heating of metal and metallic composites," *Journal of Materials Science Letters,* vol. 20, pp. 1561-1563, 2001.

[21] J. P. Cheng, R. Roy, and D. Agrawal, "Radically different effects on materials by separated microwave electric and magnetic fields," *Materials Research Innovations,* vol. 5, pp. 170-177, Feb 2002.

[22] R. Roy, R. Peelamedu, C. Grimes, J. P. Cheng, and D. Agrawal, "Major phase transformations and magnetic property changes caused by electromagnetic fields at microwave frequencies," *Journal of Materials Research,* vol. 17, pp. 3008-3011, Dec 2002.

[23] C. B. Whitehead, S. Ozkar, and R. G. Finke, "LaMer's 1950 Model for Particle Formation of Instantaneous Nucleation and Diffusion-Controlled Growth: A Historical Look at the Model's Origins, Assumptions, Equations, and Underlying Sulfur Sol Formation Kinetics Data," *Chemistry of Materials,* vol. 31, pp. 7116-7132, Sep 24 2019.

[24] M. A. Watzky and R. G. Finke, "Nanocluster size-control and "magic number" investigations, experimental tests of the "living-metal polymer" concept and of mechanism-based size-control predictions leading to the syntheses of iridium(0) nanoclusters centering about four sequential magic numbers," *Chemistry of Materials,* vol. 9, pp. 3083-3095, Dec 1997.

[25] I. M. Lifshitz and V. V. Slyozov, "The kinetics of precipitation from supersaturated solid solutions," *Journal of Physics and Chemistry of Solids,* vol. 19, pp. 35-50, 1961/04/01/ 1961.

[26] J. A. Marqusee and J. Ross, "Kinetics of Phase-Transitions - Theory of Ostwald Ripening," *Journal of Chemical Physics,* vol. 79, pp. 373-378, 1983.

[27] S. C. Erwin, L. J. Zu, M. I. Haftel, A. L. Efros, T. A. Kennedy, and D. J. Norris, "Doping semiconductor nanocrystals," *Nature,* vol. 436, pp. 91-94, Jul 7 2005.

[28] N. Razgoniaeva, P. Moroz, M. R. Yang, D. S. Budkina, H. Eckard, M. Augspurger, *et al.*, "One-Dimensional Carrier Confinement in "Giant" CdS/CdSe Excitonic Nanoshells," *Journal of the American Chemical Society,* vol. 139, pp. 7815-7822, Jun 14 2017.

[29] D. J. Norris, A. L. Efros, and S. C. Erwin, "Doped nanocrystals," *Science,* vol. 319, pp. 1776-1779, Mar 28 2008.

[30] S. Gilje, S. Han, M. Wang, K. L. Wang, and R. B. Kaner, "A chemical route to graphene for device applications," *Nano Letters,* vol. 7, pp. 3394-3398, Nov 2007.

[31] C. J. Barrows, V. A. Vlaskin, and D. R. Gamelin, "Absorption and Magnetic Circular Dichroism Analyses of Giant Zeeman Splittings in Diffusion-Doped Colloidal Cd1-xMnxSe Quantum Dots," *Journal of Physical Chemistry Letters,* vol. 6, pp. 3076-3081, Aug 6 2015.

[32] R. Beaulac, L. Schneider, P. I. Archer, G. Bacher, and D. R. Gamelin, "Light-Induced Spontaneous Magnetization in Doped Colloidal

Quantum Dots," *Science,* vol. 325, pp. 973-976, Aug 21 2009.

[33] F. V. Mikulec, M. Kuno, M. Bennati, D. A. Hall, R. G. Griffin, and M. G. Bawendi, "Organometallic synthesis and spectroscopic characterization of manganese-doped CdSe nanocrystals," *Journal of the American Chemical Society,* vol. 122, pp. 2532-2540, Mar 22 2000.

[34] C. J. Barrows, P. Chakraboriy, L. M. Kornowske, and D. R. Gamelin, "Tuning Equilibrium Compositions in Colloidal Cd1-xMnxSe Nanocrystals Using Diffusion Doping and Cation Exchange," *Acs Nano,* vol. 10, pp. 910-918, Jan 2016.

[35] C. B. Murray, D. J. Norris, and M. G. Bawendi, "Synthesis and Characterization of Nearly Monodisperse Cde (E = S, Se, Te) Semiconductor Nanocrystallites," *Journal of the American Chemical Society,* vol. 115, pp. 8706-8715, Sep 22 1993.

[36] N. Pradhan, D. M. Battaglia, Y. C. Liu, and X. G. Peng, "Efficient, stable, small, and water-soluble doped ZnSe nanocrystal emitters as non-cadmium biomedical labels," *Nano Letters,* vol. 7, pp. 312-317, Feb 2007.

[37] M. Makkar and R. Viswanatha, "Frontier challenges in doping quantum dots: synthesis and characterization," *Rsc Advances,* vol. 8, pp. 22103-22112, 2018.

[38] M. van Schilfgaarde and O. N. Mryasov, "Anomalous exchange interactions in III-V dilute magnetic semiconductors," *Physical Review B,* vol. 63, Jun 15 2001.

[39] N. S. Karan, S. Sarkar, D. D. Sarma, P. Kundu, N. Ravishankar, and N. Pradhan, "Thermally Controlled Cyclic Insertion/Ejection of Dopant Ions and Reversible Zinc Blende/Wurtzite Phase Changes in ZnS Nanostructures," *Journal of the American Chemical Society,* vol. 133, pp. 1666-1669, Feb 16 2011.

[40] D. A. Chen, R. Viswanatha, G. L. Ong, R. G. Xie, M. Balasubramaninan, and X. G. Peng, "Temperature Dependence of "Elementary Processes" in Doping Semiconductor Nanocrystals," *Journal of the American Chemical Society,* vol. 131, pp. 9333-9339, Jul 8 2009.

[41] V. Chikan, "Challenges and Prospects of Electronic Doping of Colloidal Quantum Dots: Case Study of CdSe," *Journal of Physical Chemistry Letters,* vol. 2, pp. 2783-2789, Nov 3 2011.

[42] L. G. Gutsev, N. S. Dalal, and G. L. Gutsev, "Structure and Properties of a (CdSe)(6)@(CdSe)(30) Cluster Doped with Mn Atoms," *Journal of Physical Chemistry C,* vol. 119, pp. 6261-6277, Mar 19 2015.

[43] M. Molaei, A. R. Khezripour, and M. Karimipour, "Synthesis of ZnSe nanocrystals (NCs) using a rapid microwave irradiation method and investigation of the effect of copper (Cu) doping on the optical properties," *Applied Surface Science,* vol. 317, pp. 236-240, Oct 30 2014.

[44] S. Joicy, R. Saravanan, D. Prabhu, N. Ponpandian, and P. Thangadurai, "Mn2+ ion influenced optical and photocatalytic behaviour of Mn-ZnS quantum dots prepared by a microwave assisted technique," *Rsc Advances,* vol. 4, pp. 44592-44599, 2014.

[45] D. Zhu, W. Li, H. M. Wen, Q. Chen, L. Ma, and Y. Hu, "Microwave-assisted aqueous synthesis of Mn-doped ZnS quantum dots and their room-temperature phosphorescence detection of indapamide," *Analytical Methods,* vol. 6, pp. 7489-7495, 2014.

[46] J. Zhang, Q. H. Chen, W. L. Zhang, S. L. Mei, L. J. He, J. T. Zhu, *et al.*,

"Microwave-assisted aqueous synthesis of transition metal ions doped ZnSe/ZnS core/shell quantum dots with tunable white-light emission," *Applied Surface Science*, vol. 351, pp. 655-661, Oct 1 2015.

[47] M. Shkir, M. T. Khan, I. M. Ashraf, S. AlFaify, A. M. El-Toni, A. Aldalbahi, *et al.*, "Rapid microwave-assisted synthesis of Ag-doped PbS nanoparticles for optoelectronic applications," *Ceramics International*, vol. 45, pp. 21975-21985, Dec 1 2019.

[48] M. Shkir, Z. R. Khan, K. V. Chandekar, T. Alshahrani, A. Kumar, and S. AlFaify, "A facile microwave synthesis of Cr-doped CdS QDs and investigation of their physical properties for optoelectronic applications," *Applied Nanoscience*, vol. 10, pp. 3973-3985, Oct 2020.

[49] A. Kar and A. Patra, "Optical and Electrical Properties of Eu3+-Doped SnO2 Nanocrystals," *Journal of Physical Chemistry C*, vol. 113, pp. 4375-4380, Mar 19 2009.

[50] T. Jamatia, D. Skoda, P. Urbanek, J. Sevcik, J. Maslik, L. Munster, *et al.*, "Microwave-assisted synthesis of FexZn1-xO nanoparticles for use in MEH-PPV nanocomposites and their application in polymer light-emitting diodes," *Journal of Materials Science-Materials in Electronics*, vol. 30, pp. 11269-11281, Jun 2019.

[51] S. G. Menon, A. K. Kunti, D. E. Motaung, and H. C. Swart, "A new recipe for the rapid microwave synthesis of high quantum yield Mn2+-doped ZnGa2O4 phosphors for potential forensic applications," *New Journal of Chemistry*, vol. 43, pp. 17446-17456, Nov 28 2019.

[52] A. T. Dhiwahar, M. Sundararajan, P. Sakthivel, C. S. Dash, and S. Yuvaraj, "Microwave-assisted combustion synthesis of pure and zinc-doped copper ferrite nanoparticles: Structural, morphological, optical, vibrational, and magnetic behavior," *Journal of Physics and Chemistry of Solids*, vol. 138, Mar 2020.

[53] T. S. Nirmala, N. Iyandurai, S. Yuvaraj, and M. Sundararajan, "Effect of Cu2+ ions on structural, morphological, optical and magnetic behaviors of ZnAl2O4 spinel," *Materials Research Express*, vol. 7, Apr 2020.

[54] S. Ramachandran, C. S. Dash, A. Thamilselvan, S. Kalpana, and M. Sundararajan, "Rapid Synthesis and Characterization of Pure and Cobalt Doped Zinc Aluminate Nanoparticles via Microwave Assisted Combustion Method," *Journal of Nanoscience and Nanotechnology*, vol. 20, pp. 2382-2388, Apr 2020.

[55] M. Jahdi, S. B. Mishra, E. N. Nxumalo, S. D. Mhlanga, and A. K. Mishra, "Synergistic effects of sodium fluoride (NaF) on the crystallinity and band gap of Fe-doped TiO2 developed via microwave-assisted hydrothermal treatment," *Optical Materials*, vol. 104, Jun 2020.

[56] L. E. Mathevula, B. M. Mothudi, and M. S. Dhlamini, "Effect of Er3+ on structural and optical properties of microwave synthesized alpha-Fe2O3 nanoparticles," *Physica B-Condensed Matter*, vol. 578, Feb 1 2020.

[57] R. O. Yathisha, Y. A. Nayaka, H. T. Purushothama, P. Manjunatha, K. V. Basavarajappa, and M. M. Vinay, "Investigation the influence of Zn2+ doping on the photovoltaic properties (DSSCs) of MgO nanoparticles," *Journal of Molecular Structure*, vol. 1217, Oct 5 2020.

[58] S. Tek, B. A. Vincent, L. C. Mimun, A. Ponce, and K. L. Nash, "Atomic-Scale Structural Analysis of Homoepitaxial LaF3:Yb,Tm Core-Shell Upconversion Nanoparticles Synthesized through a

Microwave Route," *Crystal Growth & Design,* vol. 20, pp. 2153-2163, Apr 1 2020.

[59] W. F. Chen, L. F. Yan, and P. R. Bangal, "Preparation of graphene by the rapid and mild thermal reduction of graphene oxide induced by microwaves," *Carbon,* vol. 48, pp. 1146-1152, Apr 2010.

[60] P. Tang, G. Hu, Y. J. Gao, W. J. Li, S. Y. Yao, Z. Y. Liu, *et al.*, "The microwave adsorption behavior and microwave-assisted heteroatoms doping of graphene-based nano-carbon materials," *Scientific Reports,* vol. 4, Aug 11 2014.

[61] K. Chen, C. L. Wang, D. Ma, W. X. Huang, and X. H. Bao, "Graphitic carbon nanostructures via a facile microwave-induced solid-state process," *Chemical Communications,* pp. 2765-2767, 2008.

[62] C. L. Wang, D. Ma, and X. H. Bao, "Transformation of Biomass into Porous Graphitic Carbon Nanostructures by Microwave Irradiation," *Journal of Physical Chemistry C,* vol. 112, pp. 17596-17602, Nov 13 2008.

[63] K. H. Lee, J. Oh, J. G. Son, H. Kim, and S. S. Lee, "Nitrogen-Doped Graphene Nanosheets from Bulk Graphite using Microwave Irradiation," *Acs Applied Materials & Interfaces,* vol. 6, pp. 6361-6368, May 14 2014.

[64] J. A. Gerbec, D. Magana, A. Washington, and G. F. Strouse, "Microwave-enhanced reaction rates for nanoparticle synthesis," *Journal of the American Chemical Society,* vol. 127, pp. 15791-15800, Nov 16 2005.

[65] S. Y. Gu, C. T. Hsieh, Y. A. Gandomi, J. K. Chang, J. Li, J. L. Li, *et al.*, "Microwave growth and tunable photoluminescence of nitrogen-doped graphene and carbon nitride quantum dots," *Journal of Materials Chemistry C,* vol. 7, pp. 5468-5476, May 14 2019.

[66] T. H. Chiang and J. F. Wager, "Electronic Conduction Mechanisms in Insulators," *Ieee Transactions on Electron Devices,* vol. 65, pp. 223-230, Jan 2018.

Chapter 5

Influence of the Microwaves on the Sol-Gel Syntheses and on the Properties of the Resulting Oxide Nanostructures

Luminita Predoană, Dániel Attila Karajz, Vincent Otieno Odhiambo, Irina Stanciu, Imre M. Szilágyi, György Pokol and Maria Zaharescu

Abstract

Among the chemical methods in the liquid phase, the sol–gel technique is a versatile and efficient method for pure or doped metal oxide films or powders preparation, showing some advantages over other preparation techniques (high homogeneity, the possibility to introducing dopants in large amount, low processing temperature and control over the stoichiometry). Combining the sol–gel (SG)method with the effect of ultrasounds(US) or microwaves (MW) leads to improving the sol–gel procedure. The microwave-assisted sol–gel method is most frequently used for obtaining nanocrystalline, monodispersed oxide nanoparticles, or to transform amorphous gels into well-crystallized nanopowders. Less studied is the influence of the microwaves on the sol–gel reactions in solutions. The benefit of using microwave-assisted sol–gel preparation highly depends on the reagents used and on the composition of the studied systems. In the present chapter, results on the influence of the microwaves on the chemical reactions that take place during the sol–gel synthesis and on the properties of the resulted samples are discussed.

Keywords: sol–gel method, microwaves, reactions in solutions, oxide nanostructures, properties

1. Introduction

Although heavily exploited in recent decades, the domain of oxide nanostructures remains of interest to researchers throughout the world. This is because that the shapes and sizes of oxide nanomaterials greatly influence their properties, which is reflected in their use in the most diverse fields [1, 2]. Oxide nanostructures have applications in catalysis, energy storage, environmental decontamination, microelectronics, medical technology, ceramics, cosmetics, and so on [3–5].

Among the most studied branches of nanostructures are metal oxides, with representatives such as TiO_2, ZnO, CuO, Fe_3O_4, WO_3, Cr_2O_3, Co_3O_4 [6].

The structure, morphology, and properties of the oxide nanostructures depend significantly on the obtaining method. A large number of available synthesis methods underlies the continuous interest in obtaining oxide nanostructures that can be used successfully in specific areas [1, 7]. However, most of these methods are limited due to the use of toxic reagents, high processing temperatures, high vacuum, expensive equipment, or long reaction times [8, 9].

Although physical methods have the advantage of high reproducibility, chemical methods in the liquid phase are more often used to obtain oxide nanostructures due to their advantages, such as low production temperature, homogeneous mixing of precursors at the molecular scale, design and control of the physico-chemical properties of final products, depending on the precursors, and the experimental conditions used [10, 11].

Among the various chemical procedures, the sol–gel method gained increasing importance in the field of materials science because it is cheap, simple, allows the introduction of dopants in large quantities, ensures high purity, and homogeneity, allows control of size, shape, and size distribution of the obtained nanomaterials [12–14].

Lately, for the preparation of functional nanomaterials, more and more attention is being paid to the use of microwave as the energy source for carrying out a chemical reaction [1, 15]. The microwave (MW) assisted sol–gel method is reported to be a simple, cheap, faster, more energy-saving, and efficient process as compared to conventional heating methods [16–18]. The use of microwaves has received increased attention in the technological field because, among other things, it reduces the reaction time from days to minutes or hours, improves the properties of synthesized nanostructures, and allows obtaining oxide nanocrystalline films on various substrates [8, 19, 20].

The improved properties of the oxide nanostructures obtained by microwaves assisted sol–gel method could be correlated to the influence of the microwaves on the chemical reactions that take place during the sol–gel synthesis, leading to the formation of different molecular species. Results on the influence of the microwaves on the chemical reactions during the sol–gel synthesis will be discussed in the present chapter.

2. General consideration on the sol-gel chemistry

Among the chemical methods in the liquid phase, the sol–gel technique is a versatile and efficient method for pure or doped metal oxide films or powders, as well as for oxide compounds preparation [21–24].

A comprehensive definition of sol–gel method assumes that the process represents the formation of an inorganic polymeric network by reactions in the solution at low temperatures. In the second step, by adequate thermal treatments, the conversion of the inorganic amorphous polymers takes place either into glasses or into crystalline materials [1, 22].

Based on the type of the precursors and the reaction medium used, two types of sol–gel processes were developed: on the bases of the alcoholic (organic) or aqueous medium.

According to Pierre [25] in both polymeric and aqueous sol–gel routes, the precursors undertake the succession of the following transformations in the presence of water:

$$\text{Hydrolysis} \rightarrow \text{polymerization} \rightarrow \text{nucleation} \rightarrow \text{growth} \qquad (1)$$

In the case of the **polymeric route**, using alkoxides (non-ionized precursors), the reactions that occur are the following:

$$M(OR)_n + xH_2O \rightarrow M(OR)_{n-x}(OH)_x + xROH \quad (2)$$

$$M(OR)_{n-x}(OH)_x + M(OR)_{n-x}(OH)_x \rightarrow M(OR)_{n-x}(OH)_{x-1}MOM(OR)_{n-x}(OH) + H_2O \quad (3)$$

$$-M - OH + HO - M- \rightarrow -M - O - M - + H_2O \quad (4)$$

The aqueous sol–gel route has also two pathways: the colloidal route [26] and the aqueous route using different chelating agents [23, 26, 27].

In the case of the **aqueous route,** which starts from colloidal solutions in aqueous medium, the following reactions take place:

$$MX_z \rightarrow M^{z+} + X^- \quad (5)$$

$$M^{z+} + nH_2O \rightarrow \left[M(OH_2)_n\right]^{z+} \quad (6)$$

$$M(H_2O)_n^{\ z+} \rightarrow M(H_2O)_{n-1}(OH)^{(z-1)} + H^+ \quad (7)$$

$$xM(H_2O)_n^{\ z+} + yOH^- + aA \rightarrow M_xO_u(OH)_{y-2u}(H_2O)_n A_a^{(xz-y-a)+} + (xn + u - n)H_2O \quad (8)$$

In the case of transition metals, it is more difficult to obtain gels, the metals having very high reactivity due to their higher electronegativity and their not satisfied coordination sphere.

To favor the gelling process, in case of the transition metals, chelating agents, as carboxylic acids or polyols, are used. A typical reaction is the following

$$M_x(H_2O)_p(OH)_y]^{z-} + mAc \rightarrow \left[M_x(H_2O)_p(OH)_{y-m}(Ac)_m\right]^{z-} + mOH^- \quad (9)$$

It is important to underline that in all mentioned cases the reactions take place simultaneously, not consequently, and they are also reversible, fact that determines a complex composition of the sol–gel solutions.

Prior to gelation, the sol–gel solution can be used to obtain thin films by using simple techniques such as dip or spin coating [23, 28].

Besides the fact that it offers the possibility of obtaining both films and powders of metal oxides at nanometric dimensions, the sol–gel method has also some advantages over other preparation techniques. Such advantages are purity, homogeneity, the possibility to introducing dopants in large quantities, ease of manufacturing, low processing temperature, control over the stoichiometry, composition, viscosity [13, 27, 29] and, in the case of thin films, easy control of thickness, as well as the ability to cover large and different type of surfaces [30, 31].

Lately, ultrasonic [32, 33] or microwave irradiation [9, 17, 18, 34–36] in sol–gel oxide nanomaterials synthesis have become methods of interest because, in addition to being cheap and environmentally friendly heating methods, offer the advantage of using shorter synthesis time, and allow the control of crystallinity, size and morphology of the resulted nanoparticles [9, 35].

3. Microwaves and their influence on the chemical reactions

Microwave radiation is a source of energy of great interest for chemical synthesis because, among other benefits, it has been observed that the use of microwaves improves the properties of obtained nanomaterials. The first reporting on the use of microwaves in a chemical synthesis dates back to 1986 [37]. Although initially microwaves have been applied in organic synthesis, lately their use has become quite widespread in obtaining inorganic products like metal oxides nanomaterials and metallic nanomaterials [38].

Microwaves are electromagnetic radiations located between infrared radiation and radio waves with frequencies between 300 MHz (100 cm) and 300 GHz (0.1 cm). For the nanomaterials synthesis in which aqueous solutions are used, 2.45 GHz frequency is commonly applied for microwave heating of the solutions, because water absorption is maximum at this value.

Subjected to a microwave field, the substances behave differently: absorb, transmit, reflect received radiation, or any combination of these three interactions. Polar substances absorb microwaves radiation, non-polar substances are transparent environments for this type of radiation, and electrical conductors reflect microwaves radiation. Therefore, microwave heating process is used for heating the materials which can absorb the microwave energy and convert it into heat especially by dipolar polarization or conduction mechanism [1, 39]. The interactions of polar molecules and ions with the electromagnetic field have already been described by many researchers. Shortly, the collisions resulting from the rotation of the dipoles during polarization and the load carriers during conduction give energy to the atoms and molecules from the solution in the form of heat [38, 39].

While conventional heating methods are slow enough and the heat transfer from the surface to the inner material or solution, producing non-homogeneous heating, microwave heating is done quickly because microwaves can penetrate the materials to a depth that depends on the dielectric properties of the material, heating them homogeneously [38]. Consequently, microwave heating can have certain benefits over conventional heating, like faster reaction, higher reproducibility, enhancement of product quality. It is instantaneous, with no heat dissipation effects, and advantageous for selective dielectric heating, as a result of the dielectric constant difference between the solvent and reactant [40].

In sol–gel synthesis, due to rapid and direct heating of the sample with microwave radiation, the instantaneous decomposition of the precursors and the obtaining of a supersaturated solution occur. In this way, the conditions for obtaining monodispersed nanoparticles (rapid and short nucleation in a supersaturated solution) can be obtained experimentally. At the same time, the *in-situ* approach of conversion of energy results in a minimized thermal gradient due to the fast heating rate consequently is providing perfect conditions for the uniform growth of nanocrystals [31, 41].

More, in the case of sol–gel synthesis using organic solvents, characterized by slow kinetics, microwave heating is an optimal method of increasing the rate of reaction [41].

From the research carried out so far, it has been observed that, by combining the sol–gel method with the microwave heating, the properties of the obtained oxide nanostructures are improved [9, 34].

Because the presence of MW, the interaction of the electromagnetic field with each molecule in the solution differs during the hydrolysis-condensation process, we can expect the formation of different molecular species as compared to the classical sol–gel synthesis.

4. Oxide nanostructures obtained by MW assisted sol–gel method

Up to now, there have been several reports regarding the synthesis of metal oxide nanomaterials by microwave-assisted sol–gel method. However, many of them have been performed using domestic microwave ovens, in which the reaction conditions cannot be accurately measured, making the experiments difficult to be reproduced.

According to the literature data, the MW irradiation in the sol–gel synthesis was used, most frequently, for precipitation of nanocrystalline metal oxides, for thermal treatment of amorphous oxide nanopowders as well as for drying and thermally treatment of the oxide films [36].

Less attention was given to study the reactions that take place in the sol–gel solutions during MW irradiation [42–45].

4.1 Pure and doped oxide nanostructures

A large number of oxides were prepared by sol–gel and microwave assisted sol–gel methods. Using MW irradiations of the solutions, preparation of several oxides were mentioned in the literature data, as MgO [46], RuO_2 [47], ZnO [16], ZrO_2 [48], WO_3 [49], SiO_2 [50], TiO_2 [35, 51]. The power of the used microwaves ranged from 140 W [51] to 850 W [47].

Among them, considerable interest is given to pure and doped TiO_2. The doping of TiO_2 was realized with a high number of elements, such as Cr [13], Ag [52], Au, Pt [14, 53], Sn-Cu-Ni [54], Fe, Pt, Pd [51] and V [55]. Doping TiO_2 with different elements the properties of the resulted nanostructures are improved, while using microwave assisted preparation, supplementary improvement was also observed.

Our studies regarding the influence of the microwaves on the reactions in the sol–gel solutions were published by Predoana et al. [42] in the case of TiO_2 and V-doped TiO_2 nanostructures.

The use of vanadium as a doping agent has a beneficial influence on the TiO_2 properties: it can reduce the band gap energy, enhance the absorption of visible light and increase the specific surface area of the powder. The mentioned properties are reflected mainly in its photocatalytic activity, previously presented by Huang et al. [55].

In our studies, the reagents used in the synthesis were titanium(IV) ethoxide $Ti(OC_2H_5)_4$ in the case of TiO_2, as well as, titanium(IV) ethoxide $Ti(OC_2H_5)_4$ and vanadylacetylacetonate VO(AcAc), for V-doped TiO_2. In both cases, ethanol C_2H_5OH as a solvent, 2,4 pentanedione (AcAc), as a chelating agent, and nitric acid HNO_3 as catalyst were used.

By the classical sol–gel method the reagents were mixed for 2 hours at room temperature. By the microwave-assisted sol–gel method, the same mixture was exposed for 5 min at 300 W and a frequency of 2.45 GHz.

The first important result of using the microwave-assisted sol–gel method is the significantly increasing of the stability of the prepared solutions against gelation, having a great advantage for multilayer film deposition. This effect was assigned to the formation of different molecular species.

The solutions were used for obtaining thin films and the resulted gels were investigated for their structural and morphological properties.

In our studies for **TiO_2 samples** synthesized by sol–gel and microwave-assisted sol–gel methods, the TG/DTG/DTA curves corresponding to the decomposition of the obtained gels are presented in **Figure 1**.

It could be noticed that the thermal decomposition of the gels is not essentially influenced by the method of preparation. Only a small increase of the thermal effect at 195° C is observed for the TiO_2 sample obtained by MW assisted sol–gel method. The fact could be explained by the positive influence of microwaves on the formation of the molecular species that decompose at the mentioned temperature.

Based on the TG/DTG/DTA results, the samples prepared by both methods were thermally treated at 450°C for 1 h. By X-ray diffraction of the samples thermally treated at this temperature only anatase phase was detected (according to

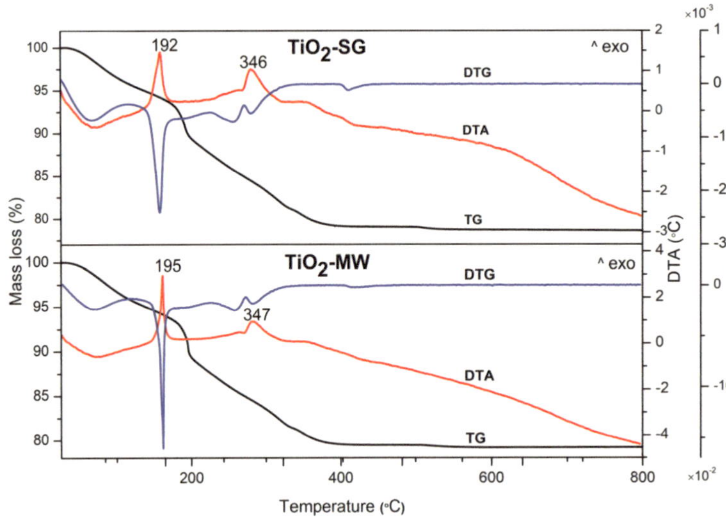

Figure 1.
TG/DTG/DTA curves of the TiO_2 samples obtained by SG and MW methods [42] (Reproduced with the permission of Springer Nature).

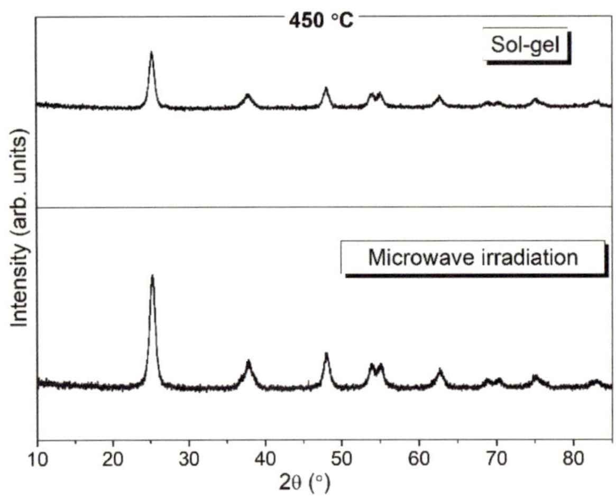

Figure 2.
The XRD patterns of the TiO_2 samples obtained by SG and MW-assisted SG methods thermally treated at 450°C.

JCPDS card no. 21–1272), but a higher crystallinity is noticed in the case of sample obtained by MW-assisted procedure (**Figure 2**).

In the case of **the V-doped TiO_2** the TG/DTG/DTA measurements in the air are presented in **Figure 3** for the gel containing 2 mol% V. In this case, increased thermal stability and a more complex decomposition of the gels obtained by the microwave-assisted sol–gel method is observed.

Confirmations of the TG/DTG/DTA results on the gels with 2 mol% V were obtained by Differential Scanning Calorimetry (DSC). The obtained DSC curves are presented in **Figure 4**.

According to the DSC results, the thermal stability of the gel obtained from the solution prepared in the presence of microwaves, is significantly higher (with about 100°C), as compared with the gel with similar compositions, but obtained by the classical sol–gel method.

At the same time, the number and temperatures of the thermal effects are different in the two discussed cases underlying the different compositions of the gels obtained in the presence or the absence of the microwaves.

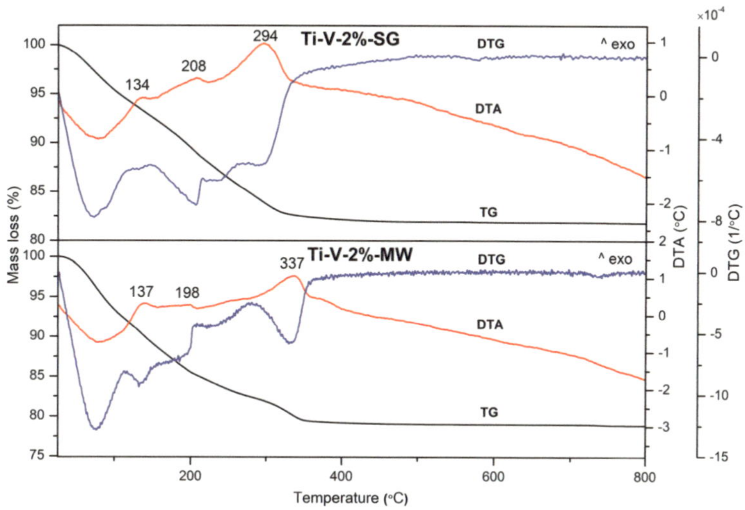

Figure 3.
TGA/DTG/DTA curves of the V-doped TiO2 samples obtained by SG and MW-assisted SG methods [42] (Reproduced with the permission of Springer Nature).

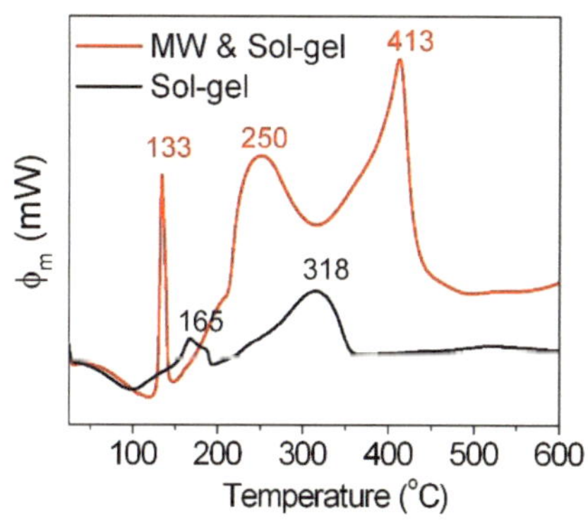

Figure 4.
DCS curves of the V-doped TiO_2 obtained by SG and MW-assisted SG methods [43] (Reproduced with the permission from Springer Nature).

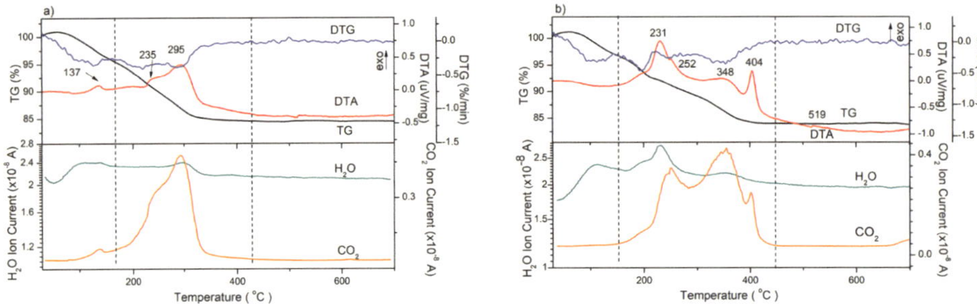

Figure 5.
TG/DTG/DTA/EGA curves of V-doped TiO_2 obtained by (a) SG and (b) MW-assisted SG methods [42] (Reproduced with the permission of Springer Nature).

The TG/DTG/DTA/EGA measurements, presented in **Figure 5**, have confirmed, once more, the results discussed above, regarding the different thermal behavior of the gels obtained by the microwave-assisted sol–gel method.

In the case of the microwave-assisted sol–gel method the same gasses are evolved, namely H_2O and CO_2, but a more complex thermal decomposition is observed, with different ratios among the two mentioned gases at the different temperatures. This result is assigned to the higher number of molecular species present in the gel, having different chemical composition and different thermal stability.

By X-ray diffraction of the V-doped TiO_2 with 2 mol% V samples thermally treated at 450°C (**Figure 6**) only anatase phase was detected (according to JCPDS card no. 21–1272). As in the case of un-doped TiO_2, a higher crystallinity is noticed in the case of samples obtained by MW assisted procedure.

Before gelation, the solutions prepared in the presence and in the absence of MWs were used for thin film deposition by dip-coating on glass substrates [43].

In our studies for **the TiO_2 films** obtained by the sol–gel method, the SEM micrographs show surface cavities that were not observed in the case of microwaves-assisted sol–gel films (**Figure 7a** and **c**).

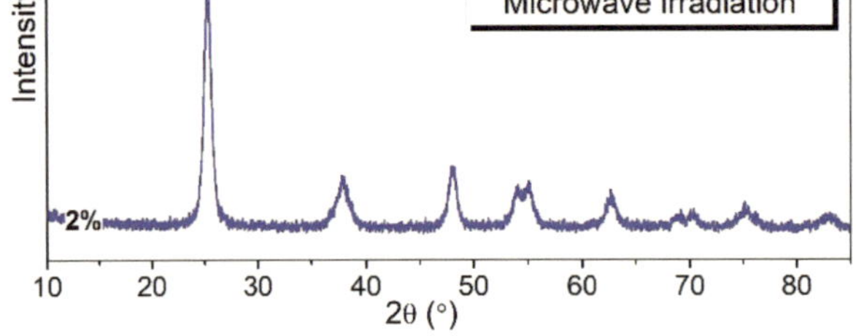

Figure 6.
The XRD patterns of the V-doped TiO_2 samples obtained by SG and MW-assisted SG methods, thermally treated at 450°C.

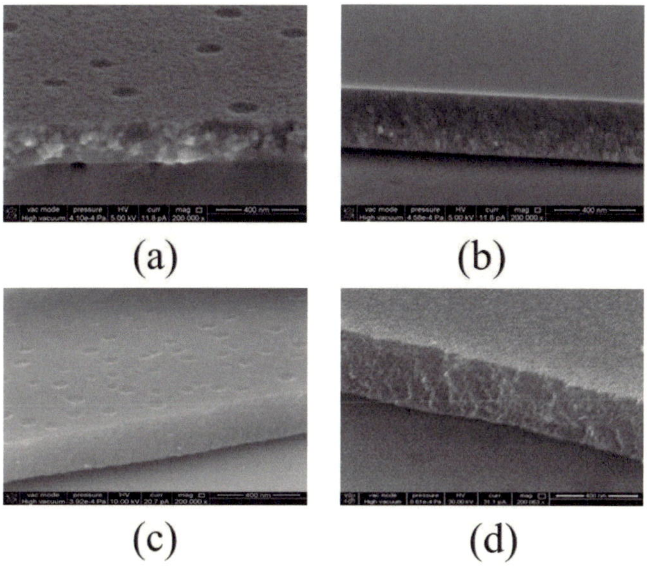

Figure 7.
SEM micrographs showing the film cross-section for samples (a) $(TiO_2)_{SG}$*; (b)* $(TiO_2)_{MW}$*; (c)* $(V\text{-doped}TiO_2)_{SG}$*; (d)* $(V\text{-doped}TiO_2)_{MW}$ *[43].*

The sol–gel TiO_2 based films present also a similar variation of the morphology according to the method of preparation. A more dense and homogeneous aspect is observed in the film obtained in the presence of microwaves (**Figure 7b** and **d**).

Thickness values are around 200 nm both for TiO_2 and V-doped TiO_2 films, but slightly higher in the case of the films obtained from microwave-assisted sol–gel solutions.

The transmission spectra of obtained films are presented in **Figure 8** show optical transmittance values mainly over 80% in the visible range.

To explain the differences induced by the microwave-assisted sol–gel process on the properties of the resulted films, their influence on the starting solution, and the evolution of the sol–gel process, should be taken into consideration. Based on the results obtained up to now, it could be assumed that in the presence of microwaves, different and more stable molecular species are formed as compared to the classical sol–gel method and this a fact influences the properties of the resulted films.

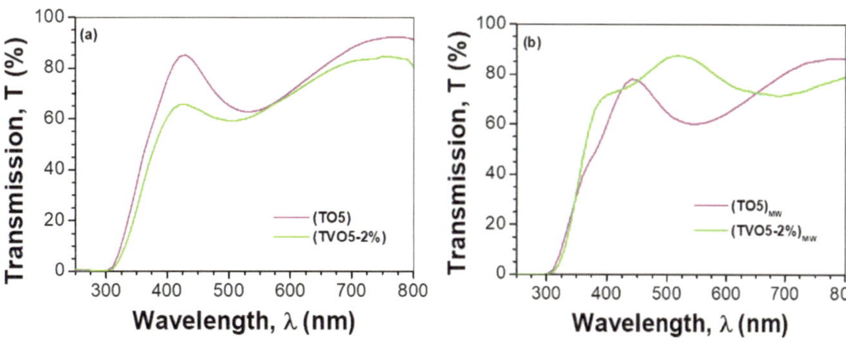

Figure 8.
Optical transmission of the TiO2 and V-doped TiO2 films obtained by (a) SG and (b) MW-assisted SG methods.

It was also observed that the effect of microwaves on the properties of the resulted materials is higher in the case of V-doped TiO_2 samples, fact that could be correlated to an enhancement of the reactions between Ti and V reagents during the sol–gel process in the presence of the microwaves.

As presented in the several references, **WO_3 based nanomaterials** are widely investigated in the field of electrochromic devices [56], gas sensing [57], and photocatalysis [58] in different morphologies and structures. Even though the sol–gel process has a long past and is an intensely researched method [59] the literature of sol–gel preparation of WO_3 using microwave-assistance is scarce. The following articles are all from the 2010s so further researches are to be expected.

Different nanostructures were prepared by microwave assisted sol–gel method with sodium tungstate as a precursor material by Kharade et al. [60–62]. The research group synthesizes various nanoparticles and nanofilms for electrochromic purposes. In 2012 WO_3 nanofilms were deposited on the FTO substrate, which was the first time used MW-assisted two-step process. In the first step, the preparation of the gel was conducted with microwave assistance, then in the second step, the deposition of the thin film occurred by a chemical growth set up. Scanning electron microscope (SEM) showed that the surface is coated with petal-like WO_3 nanodisks with dimensions of 450–600 nm length, 350–400 nm width, and 20–35 nm thickness. The X-ray diffraction (XRD) analysis (**Figure 9**) points out that WO_3 is in the hexagonal crystal form. Narrow and intense XRD peaks indicate that the material has good crystallinity and calculations determined that the crystal size is 71 nm, which is comparable to samples made by the regular sol–gel method [63]. X-ray photoelectron spectroscopy (XPS) revealed that the W:O ratio is non-stoichiometric(2.89). Electrochromic capabilities were determined with different electroanalytical methods [60]. Comparing this to a regular sol–gel method shows that the morphology of the surface, namely the platelet like nanodisks is nearly the same with a small difference in size (regular sol–gel platelets: 10–30 nm thick and few hundred nm lengths and width). However, to achieve the same crystallinity a 500°C annealing process is required for the regular sol–gel method, in contrast to the 150°C drying of the MW-assisted sol–gel method [64].

The same hexagonal WO_3 thin film was synthesized and its electrochromic properties were enhanced with different amounts of Ag nanoparticles [61]. The microwave-assisted sol–gel method was also used to produce WO_3/MoO_3 mixed

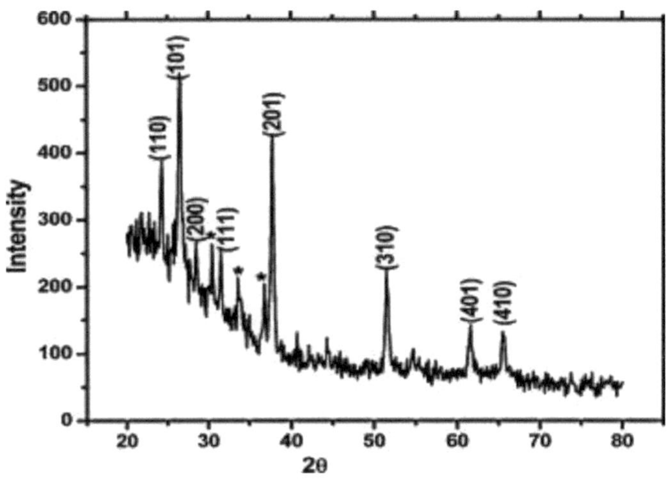

Figure 9.
X-ray diffractogram of the WO_3 thin film [60] (Copyright (2012), with permission from Elsevier).

oxide thin films. First, the WO_3 layer were produced with the two-step method explained earlier, then MoO_3 was deposited with vacuum evaporation [62].

Hilaire et al. [49] prepared WO_3 nanoparticles using a nonaqueous microwave-assisted sol–gel method for photoanodes. The synthesized nanoparticles were analyzed with FT-IR, which showed that no organic contaminant remained on the surface of the particles, but a weight indicates that there are a 4.4% water and organic residue after 800°C heating. XRD studies confirm the monoclinic crystalline structure of the WO_3 nanoparticles.

Transmission electron microscopy (TEM) showed that the platelets like WO_3 nanoparticles size is 20–40 nm and thickness of 3 nm. Moreover, TEM measurements indicate that the WO_3 platelets face having the crystalline orientation of [0 0 2]. The WO_3 nanoparticles were used for the production of photoanodes, which was proven to be an efficient method for water splitting. The comparison of this result with another nonaqueous regular sol–gel method shows that the morphology of the particles differs, but this can be caused by the usage of a different solvent (dicarboxylic acid) and modifier (polyethylene glycol).

The regular method resulted in larger (58 nm) rod-like nanoparticles. The case of the WO_3 particle's crystallinity is similar to the thin layer's: without after annealing process, the MW assisted method provides better crystallinity [65].

It was also established [66] that microwave heating is more convenient than resistive heating to fabricate WO_3 nanoparticles with high specific surfaces and very small particle sizes also in the case of hydrothermal method of preparation. In our studies [67, 68] hexagonal structured WO_3 nanoparticles and wires were prepared using MW assisted hydrothermal process. SEM images are presented in the **Figure 10**.

The Au decorated h-WO_3 nanowires were prepared for photocatalysis. The pre-decorated WO_3 nanowires showed crystallinity and were composed of W and O only. The morphology also differs from nanodisks, the hydrothermally produced WO_3 took the form of nanorods with 10 μm length and 10 nm diameter.

Nevertheless, the Au decorated nanowires showed great photocatalytic activities. Nanowires and nanoparticles coated with TiO_2 using ALD were also

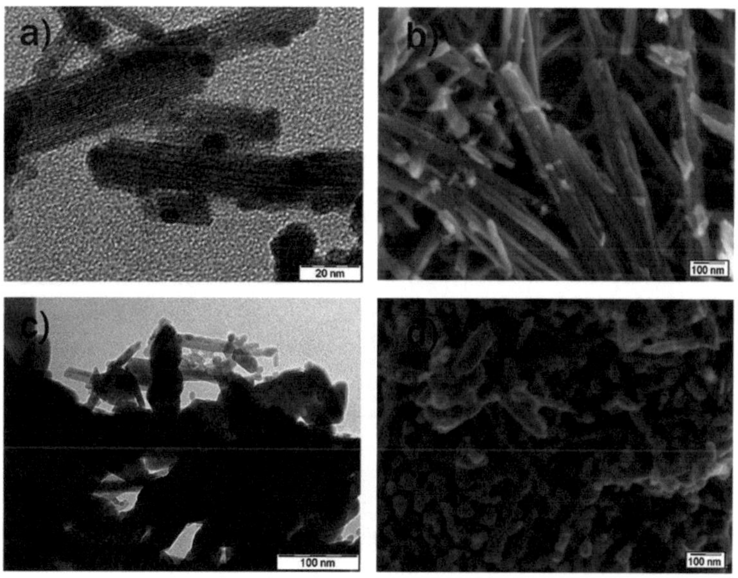

Figure 10.
SEM images of the (a,b) hexagonal WO_3 nanowire coated with TiO_2 and (c, d) monoclinic WO_3 nanoparticle coated with TiO_2 [67] (Reprinted with permission from [67] copyright from RSC Advances).

synthesized, but the characteristics of the non-coated samples were done. Hexagonal and monoclinic nanoparticles were prepared using controlled annealing of the samples.

Similarly, further annealing is needed to reach a comparable crystallinity, but for the monoclinic structure it's obligatory. The size of the crystals was 50–70 nm and 60–90 nm for hexagonal, and for the irregular shaped monoclinic WO_3 nanoparticles respectively. The hexagonal WO_3 nanowires were analogous to the earlier nanowire, several μm long and 5–10 nm diameter. The TiO_2 coated nanostructures proved to be efficient photocatalysts [67, 68].

4.2 Oxide compounds

Let us deal with the results regarding the synthesis by microwave-assisted sol–gel methods of the precursor powders for $SrCu_2O_2$ preparation.

The interest for the $SrCu_2O_2$ compound are connected to its possible applications as thermoelectric or full oxide electronic devices, solar cells, liquid-crystal displays, touchscreen, and so on [45].

Among the CuO-based p-type TCOs, Cu-Sr-O has received attention due to its wide direct band gap, and its potential use in transparent optoelectronic devices; such as light-emitting diodes, laser diodes, solar cells, display technology, and other technologies [69]. In most of the published reports, Cu based p-type TCO thin films are deposited by high vacuum processes which are costly. Some of the processes include pulsed laser deposition (PLD), reactive evaporation, magnetron sputtering, thermal co-evaporation and radio frequency [70–76]. To date, few studies have reported on the preparation of a Cu-based p-type TCO by a non-vacuum solution chemical route.

Roy et al. [77] used sol–gel and annealing methods to prepare Cu_2SrO_2 thin films. They used different oxygen pressure, annealing time, and temperature combinations to attempt to obtain phase pure Cu_2SrO_2 thin films. Copper (II) methoxide and triethanolamine were mixed in the ration 1:1. Pure Sr-metal was dissolved separately in distilled anhydrous isopropanol under argon. The Cu-solution was then mixed drop-wise into the Sr-solution while stirring. The mixture was stirred continuously for 2 hrs at room temperature. The sol was spin-coated on clean substrates with 3000 rpm for 30 s. The coated films were heated at 225°C for 2 min in the air for partial pyrolysis. This coating/heating cycle was repeated ten times to obtain films of the desired thickness of 500 nm. After deposition, the film was annealed further under controlled oxygen pressure. Different annealing procedures were used to avoid the presence of excess Cu_2O phase.

XRD analysis (**Figure 11b**) showed the films had a mixed-phase of excess Cu_2O and Cu_2SrO_2 after final reduced-oxygen pressure annealing. Films annealed at lower oxygen pressure (1.3×10^{-2} and 1.3×10^{-3} Pa) had similar phase composition and in all the three films Cu_2O formed as a secondary phase with Cu_2SrO_2. For the film annealed at the highest oxygen pressure (1.3×10^{-1} Pa), $CuSrO_2$ was observed as the amount of Cu_2SrO_2 decreased and the intensity of the Cu_2O peaks did not change.

Both SEM and TEM images (**Figure 11a** and **c**) show that two phases are present. The light-gray particles (differing sizes) in the SEM and large particles in TEM images are the Cu_2SrO_2 phases. The dark gray phase in the SEM image is a mixture of small Cu_2SrO_2 and Cu_2O particles, as confirmed by the TEM images. The SEM and TEM images reveal that the Cu_2O and Cu_2SrO_2 phases are intermingled with each other.

Ginley et al. [78] used sol–gel and annealing to prepare pure phased Cu_2SrO_2 films. Stoichiometric amounts of aqueous solutions copper formate and strontium acetate were mixed in methanol and stirred. Triethanolamine was added, the

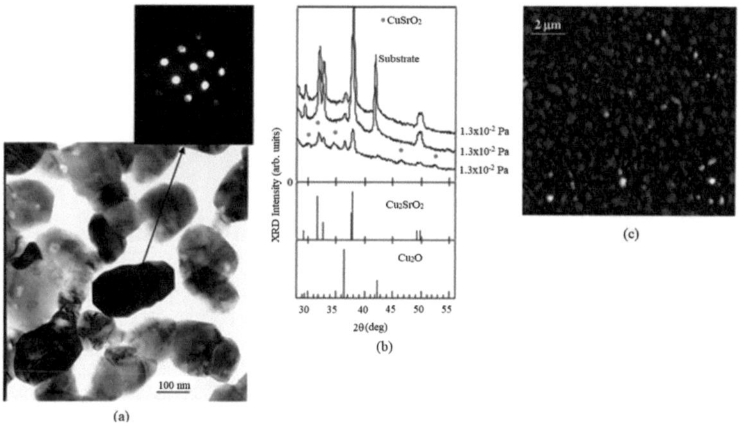

Figure 11.
(a) TEM image, (b) XRD spectra (c) SEM image of the films after final annealing at 750°C under 1.3×10^{-2} Pa oxygen pressure [77] (Reproduced by the permission of Elsevier).

mixture stirred and evaporated at 80°C to form sol which was diluted by isopropyl alcohol and spin-coated on MgO (100) substrates for 20 s, at 3000 revs per min. The resulting films were annealed at 200°C temperature for 2 min and then pyrolyzed at 500°C for 2 min. The spin-coating and pyrolysis cycles were repeated 8–10 times. After the cycles, the films were first annealed at 750°C for 30 min in air and then at 775°C under 2.7×10^{-6} Torr oxygen. The films were characterized by XRD (**Figure 12**) and FTIR and showed to be phase pure.

Predoana et al., are the first to report the synthesis of Sr-Cu-O gels by microwave (MW) assisted sol–gel methods [45]. Pure strontium acetyl acetonate ($Sr(C_5H_7O_2)_2$ and copper (II) acetyl acetonate ($Cu(C_5H_7O_2)_2$) were used as precursors for strontium and copper, respectively. The 0.25 M aqueous solutions of Sr.$(C_5H_7O_2)_2$ and $Cu(C_5H_7O_2)_2$ solution in absolute ethanol were mixed with triethanolamine, in the ratio 1:1. In the case of the sol–gel method, the starting solution was homogenized under vigorous stirring for 2 h at 80 C. For MW assisted sol–gel method, the same starting solution was homogenized by stirring and exposing to microwaves having power ~ 300 W and 2.45 GHz frequency for 5 minutes. The sol–gel and the microwave-assisted sol–gel prepared Sr-Cu-O were characterized by SEM, FTIR, XRD, and their thermal properties investigated by TG/DTA-MS in air, inert and reducing atmospheres.

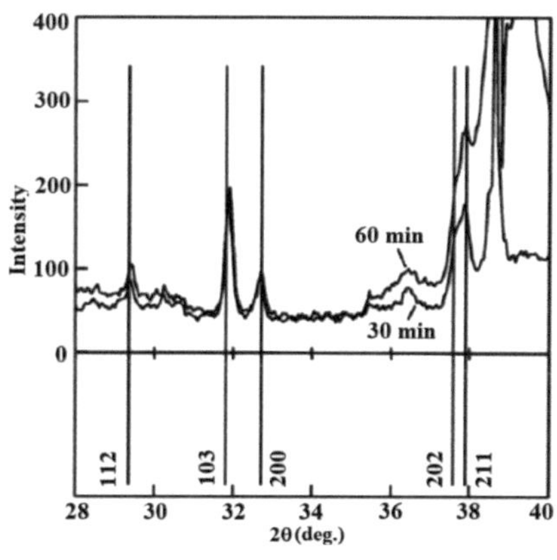

Figure 12.
XRD patterns of Cu_2SrO_2 films as a function of processing time [78].

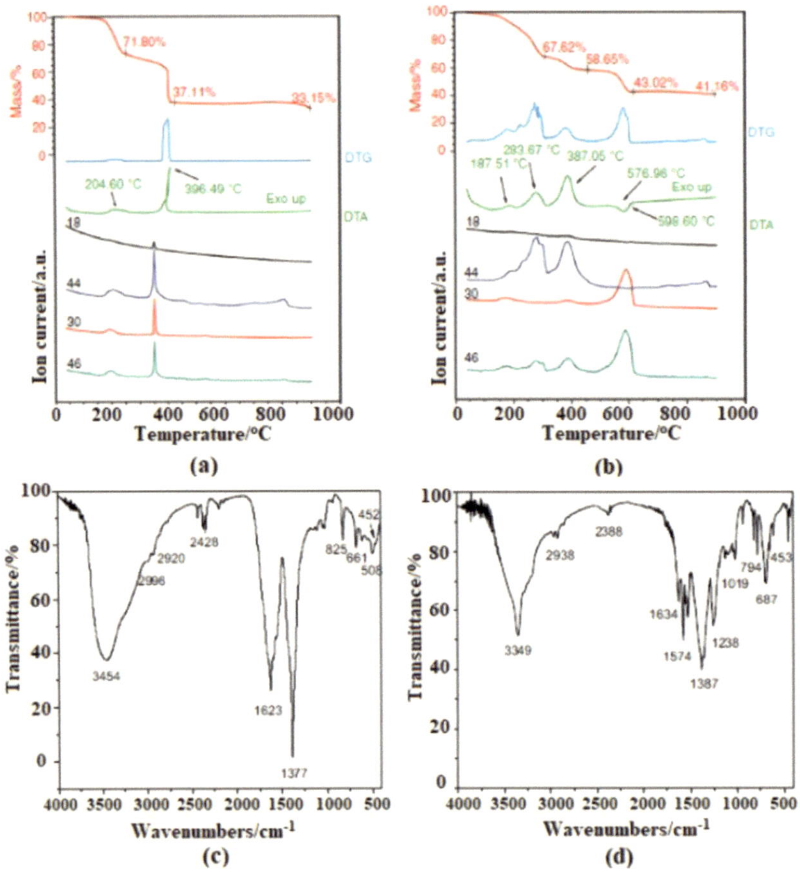

Figure 13.
Thermal decomposition in air (a) sol–gel synthesized sample, (b) MW assisted sol–gel synthesized sample, (c) FTIR spectra of sol–gel synthesized sample, (d) FTIR spectra of MW assisted sol–gel synthesized sample [45] (Reproduced by the permission of Elsevier).

In the experimental conditions presented above pieces of gels of different size, and blueish-green color were obtained for both preparation methods. The results obtained by TG/DTA-MS analysis (**Figure 13a** and **b**) of the obtained gels demonstrated the influence of MW on the sol–gel synthesis. MW treated samples had one more mass loss step when heated in air attributed to complex compositions of the resulted gels that contain a higher number of molecular species with higher thermal stability. The results were confirmed with the FTIR spectra (**Figure 13c** and **d**) showing more vibration bands for the samples prepared by the MW sol-gel method, assigned according to [79–81].

Based on the XRD patterns of the residues (**Figure 14**), the final product is composed of a mixture of phases that depend on the synthesis route and the annealing conditions.

For samples annealed in air, Sr–Cu–O phase was also present for the sol–gel synthesized sample, while the MW sample had CuO as the main component. In different atmosphere (N_2 and H_2/Ar) several compounds (Sr_2CuO_3, SrO and CuO) are present in varying amounts. Only traces of $SrCO_3$ can be detected. In all annealing atmospheres, in the case of the samples synthesized by MW-assisted sol–gel method, powders with a lower degree of crystallization is formed. This result could be attributed to the formation of a higher number of molecular species with higher thermal stability.

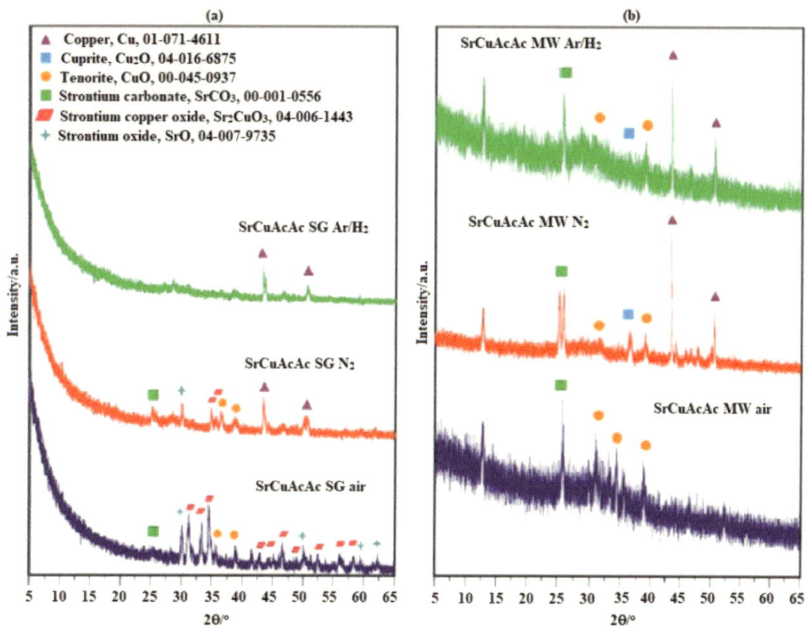

Figure 14.
(a) XRD patterns of sol–gel synthesized samples (b) MW-assisted sol–gel samples annealed at 900°C in air, N_2 and 5%H_2/95%Ar [45].

The powders prepared in the mentioned conditions are intended to be investigated as precursors for $SrCu_2O_2$ compound preparation.

The presented results are important revealing the effect of MW on the reactions that take place during the sol–gel synthesis but should be considered preliminary. Direct methods of the solutions investigations, as High-Pressure Liquid Cromatogaphy (HPLC), are underway in order to bring more information on the sol–gel chemistry in the presence and the absence of microwaves.

5. Conclusions

The interest of using microwaves in obtaining oxide nanostructures by reactions in solutions is rather high, leading to obtaining powders or films with enhanced properties.

According to the literature data, the MW irradiation in the sol–gel synthesis was used, most frequently, for precipitation of nanocrystalline metal oxides, for thermal treatment to crystallize the amorphous oxide nanopowders as well as for drying and thermally treatment of the oxide films.

However, the influence of the microwaves on the chemical reactions that take place during the sol–gel synthesis is less investigated.

Results regarding the formation of pure or doped nanostructures, as well as oxide compound, by sol–gel method in the presence or absence of microwave are presented.

The main results of the studies have shown that in all cases in the presence of microwave formation different molecular species is observed with a positive influence on the properties of the resulted nanostructure.

The advantage of using the MW-assited sol–gel method is a more shorter time of synthesis and obtaining nanostructures with improved properties.

The obtained results are of interest, but could be considered preliminary and systematic studies on the chemical processes induced by the microwaves should be continued.

Acknowledgements

This work was performed in the frame of Mobility Project "Reduced semiconductor oxides for TCO, photocatalysis and gas sensing applications", 2019– 2021, between IlieMurgulescu Institute of Physical Chemistry of the Romanian Academy, Bucharest, Romania and Research Center for Natural Sciences, Hungarian Academy of Science Research Group of the Hungarian Academy of Science at the Budapest, Hungary. An NRDI K 124212 and an NRDI TNN_16 123631 grants are acknowledged. The research within project No. VEKOP-2.3.2-16-2017-00013 was supported by the European Union and the State of Hungary, co-financed by the European Regional Development Fund. The research reported in this paper was supported by the BMENanotechnology and Materials Science TKP2020 IE grant of NKFIH Hungary (BME IE-NAT TKP2020)and Stipendium Hungaricum scholarship grant.

Author details

Luminita Predoană[1], Dániel Attila Karajz[2], Vincent Otieno Odhiambo[2], Irina Stanciu[3], Imre M. Szilágyi[2], György Pokol[2,4] and Maria Zaharescu[1*]

1 "Ilie Murgulescu" Institute of Physical Chemistry, Romanian Academy, Bucharest, Romania

2 Department of Inorganic and Analytical Chemistry, Budapest University of Technology and Economics, Hungary

3 Constanta Maritime University, Constanța, Romania

4 Research Center for Natural Sciences, Hungarian Academy of Science, Hungary

*Address all correspondence to: mzaharescu2004@yahoo.com

References

[1] Garadkar KM, Kadam AN, Park J. Microwave-Assisted Sol–Gel Synthesis of Metal Oxide Nanomaterials. In: Klein L, Aparicio M, Jitianu A, editors. Handbook of Sol-Gel Science and Technology.Processing, Characterization andApplications.2nd ed.Cham: Springer; 2016. p. 1-22. DOI: 10.1007/978-3-319-19454-7_107-1

[2] Uglov VV, Doroshevich IL, Kvasov NT, RemnevGE, Shymanski VI. On physical properties of nanoparticles: Size effect and scale of nanoobjects. Phys Status Solidi C. Current Topics in Solid State Physics.2016;13:903-907. DOI:10.1002/pssc.201600039

[3] Canas-Carrell JE, Li S, Parra AM, Shrestha B. Metal oxide nanomaterials: health and environmental effects. In: Njuguna J, Pielichowski K, Zhu H, editors. Health and Environmental Safety of Nanomaterials. 1st ed.PolymerNancomposites and Other Materials Containing Nanoparticles. WoodheadPublishing; 2014. p. 200-221. DOI: 10.1533/9780857096678.3.200

[4] Heera P, Shanmugam S. Nanoparticle characterization and application: An overview. Int J CurrMicrobiolAppl Sci. 2015;4:379-386.ISSN: 2319-7706

[5] Khan I, Saeed K, Khan I. Nanoparticles: Properties, applications and toxicities. Arab J Chem. 2019;12:908-931. DOI: 10.1016/j.arabjc.2017.05.011

[6] Dey K, Srivastana AK. Metal oxide nanomaterials: an overview. In: Srivastava AK editor. Oxide Nanostructures: Growth, Microstructures, and Properties. Pan Stanford Publishing; 2009. p. 1-77. DOI: 10.1201/b15633-2

[7] Patil KC, Hegde MS, Rattan T, Aruna ST. Chemistry of nanocrystalline oxidematerials. Combustionsynthesis, propertiesandapplications, New Jersey:World Scientific; 2008. 364 p. DOI: 10.1142/6754

[8] Kim K, Kim E, Kim Y, Park K. Characteristics of ZnO Thin Film Transistors Fabricated Using a Microwave Sol-Gel Method. Korean J Met Mater. 2014; 52:155-161. DOI:10.3365/KJMM.2014.52.2.155

[9] MudassarMuzaffar S, Naeem S, Yaseen S, Riaz S, Kayani ZN, Naseem S. Microwave assisted tuning of optical and magnetic properties of zinc oxide nanorods—efficient antibacterial and photocatalytic agent. J Sol-Gel Sci Technol. 2020;95:88-100. DOI: 10.1007/s10971-020-05290-3

[10] Cao G, Wang Y.Nanostructures and Nanomaterials: Synthesis, Properties and Applications Applications. 2nd ed. World Scientific Series in Nanoscience and Nanotechnology, vol 2;2011. 581 p.

[11] Carp O, Huisman CL, Reller A. Photoinduced reactivity of titanium dioxide. Prog Solid State Chem.2004;32:33-177. DOI: 10.1016/j.progsolidstchem.2004.08.001

[12] Darshana B, Parikh S, Shah M. Potential of Ag–Fe co-doped TiO_2 nanocomposite for solar photocatalysis of high COD pharmaceutical effluent and influencing factors. Energ. Ecol. Environ. 2020;5:344-358. DOI: 10.1007/s40974-020-00162-6

[13] Mendiola-Alvarez SY, Guzman-Mar JL, Turnes-Palomino G, Maya-AlejandroF,Hernandez-RamirezA, Hinojosa-Reyes L. UV and visible activation of Cr(III)-doped TiO_2 catalyst prepared by a microwave-assisted sol–gel method during MCPA degradation. Environ SciPollut Res. 2017;24:12673-12682. DOI 10.1007/s11356-016-8034-x

[14] Hernandez R, Rosendo Hernandez-Resendiz J, Martinez-ChavezA,Velazquez-CastilloR, Escobar-Alarcon L, Esquivel K. X-ray diffraction Rietveld structural analysis of Au–TiO_2 powders synthesized by sol–gel route coupled to microwave and sonochemistry. J Sol-Gel Sci Technol. 2020;95:239-252. DOI: 10.1007/s10971-020-05264-5

[15] Akbar A, Riaz S, Ashraf R, Naseem S. Magnetic and magnetization properties of iron oxide thin films by microwave assisted sol–gel route. J Sol-Gel Sci Technol. 2015;74:320-328. DOI: 10.1007/s10971-014-3528-9

[16] Assi N, Azar PA, Tehrani MS, Husain SW. Studies on photocatalytic performance and photodegradation kinetics of zinc oxide nanoparticles prepared by microwave-assisted sol–gel technique using ethylene glycol. J Iran Chem Soc. 2016;13:1593-1602. DOI: 10.1007/s13738-016-0875-1

[17] Parangusan H, Ponnamma D, Al-Maadeed MAA. Effect of cerium doping onthe optical and photocatalytic properties of ZnOnanoflowers. Bull Mater Sci. 2019;42:179. DOI: 10.1007/s12034-019-1865-6

[18] Falk GS, Borlaf M, Lopez-Munoz MJ, Farinas JC, Rodrigues Neto JB, Moreno R. Microwave-assisted synthesis of TiO_2 nanoparticles: photocatalytic activity of powders and thin films. J Nanopart Res. 2018; 20:23. DOI: 10.1007/s11051-018-4140-7

[19] Dinelli M, Fabbri E, Bondioli F. TiO_2–SiO_2 hard coating on polycarbonatesubstratebymicrow aveassisted sol–gel technique. J Sol-Gel SciTechnol. 2011;58:463-469. DOI: 10.1007/s10971-011-2413-z

[20] Giesz P, Celichowski G, Puchowicz D, Kaminska I, Grobelny J, Batory D, Cieslak M. Microwave-assisted TiO_2: anatase formation on cottonandviscose fabric surfaces. Cellulose. 2016;23:2143-2159. DOI: 10.1007/s10570-016-0916-z

[21] Sakka S. Sol-gel Process and Applications. In: Somiya S., editor. Handbook of Advanced Ceramics: Materials, Applications, Processing and Properties. 2nd ed. Boston: Academic Press; 2013. p. 883-910.

[22] Schmidt H. Chemistry of material preparation by the sol-gel process. J Non-Cryst Solids. 1988;100:51-64. DOI: 10.1016/0022-3093(88)9006-3

[23] Brinker CJ, Scherer GW. Sol-gel science: the physics and chemistry of sol-gel processing. 1st ed. Boston: Academic Press; 1990. 912 p.

[24] Vartanyan M, Voytovich I, Gorbunova I, Makarov N. Preparation and Structural Characterization of Complex Oxide Eutectic Precursors from Polymer–Salt Xerogels Obtained by Microwave-Assisted Drying. Materials. 2020;13:1808. DOI: 10.3390/ma13081808

[25] Pierre AC. Introductiontothe sol–gel process. Boston: Kluwer Academic; 1998. 394 p. DOI: 10. 1007/978-1-4615-5659-6

[26] Iler RK. The chemistry of silica: Solubility, Polymerization, Colloid and Surface Properties and Biochemistry. Chichester: Wiley; 1979. 896 p

[27] Livage J, Henry M, Sanchez C. Sol–gel chemistry of transition metal oxides. Prog Solid State Chem. 1988;18:259-341. DOI: 10.1016/0079-6786(88)90005-2

[28] Atkinson C, Sansom CL, Almond HJ, Shaw CP. Coatings for concentrating solar systems – a review.

Renew Sustainable Energy Rev. 2015;45:113-122. DOI: 10.1016/j.rser.2015.01.015

[29] Akpan UG, Hameed BH. The advancements in sol–gel method of doped-TiO_2 photocatalysts - a review. ApplCatal A: General. 2010;375:1-11. DOI: 10.1016/j.apcata.2009.12.023

[30] Manole V, Dobromir M, Girtan M, Mallet R, Rusu G, Luca D. Optical properties of Nb-doped TiO_2 thin films prepared by sol–gel method. Ceram Int.2013;39:4771-4776. DOI: 10.1016/j.ceramint.2012.11.066

[31] Wang Y, He Y, Lai Q, Fan M. Review of the progress in preparing nano TiO_2: An important environmental engineering material. J Environ Sci. 2014;26:2139-2177.DOI: 10.1016/j.jes.2014.09.023

[32] Babu SG, Neppolian B, AshokkumarM.Ultrasound-Assisted Synthesis of Nanoparticles for Energy and Environmental Applications. In Ashokkumar M, editor. Handbook of Ultrasonics and Sonochemistry. Singapore:Springer; 2016. p. 423-456. DOI: 10.1007/978-981-287-278-4_16

[33] Goswami YC, Kumar V, Rajaram P, Ganesan V, Malik MA, O'Brien P. Synthesis of SnO_2 nanostructures by ultrasonic-assisted sol–gel method. J Sol-Gel Sci Technol. 2014;69:617-624. DOI: 10.1007/s10971-013-3241-0

[34] Ishtiyaq MF, Bhuiyan AA, Roy U, BillahMM.Efect of the microwave-assisted thermal annealing on structural and optical properties of the sol–gel-derived ZnO thin flms. SN Applied Sciences. 2020;2:1294. DOI: 10.1007/s42452-020-3098-0

[35] Jaimy KB, Vidya K, Saraswathy HUN, Hebalkar NY, Warrier KGK. Dopant-free anatase titanium dioxide as visible-light catalyst: Facile sol–gel microwave approach. JECE. 2015;3:1277-1286. DOI: 10.1016/j.jece.2014.06.023

[36] Predoana L, Stanciu I, Zaharescu M. Metal oxide nanomaterials obtained by sol-gel and microwave assisted sol-gel method. In: Yurish SY, editor. Advances in Microelectronics: Reviews, Volume 2. Barcelona: IFSA Publishing; 2019. p. 221-250

[37] Gedye R, Smith F, Westaway K, Ali H, Baldisera L, Laberge L, RousellJ. The use of microwave ovens for rapid organic synthesis. TetrahedronLett. 1986;27:279-282. DOI: 10.1016/S0040-4039(00)83996-9

[38] Borja Diaz de Grenu, Ruth de los Reyes, Costero AM, Amoros P, Ros-Lis JV. Recent Progress of Microwave-Assisted Synthesis of Silica Materials. Nanomaterials. 2020;10:1092. DOI: 10.3390/nano10061092

[39] Sun J, Wang W, Yue Q. Review on Microwave-Matter Interaction Fundamentals and Efficient Microwave-Associated Heating Strategies. Materials. 2016;9:231. DOI: 10.3390/ma9040231

[40] Das S, Mukhopadhyay AK, Datta S, Basu D. Prospects of microwave processing: An overview. Bull Mater Sci.2009;32:1-13. DOI: 10.1007/s12034-009-0001-4

[41] Sciancalepore C, Bondioli F, Manfredini T, Gualtieri A. Quantitativephaseanalysisandmicro structurecharacterization of magnetite nanocrystalsobtainedby microwaveassisted non-hydrolytic sol–gel synthesis. MaterCharact. 2015;100:88-97. DOI: 10.1016/j.matchar.2014.12.013

[42] Stanciu I, Predoana L, PandeleCusu J, Preda S, Anastasescu M,

Vojisavljevic K, Malic B, ZaharescuM. Thermal behaviour of the TiO_2-based gels obtained by microwave-assisted sol-gel method. JTAC. 2017;130:639-651. DOI:10.1007/s10973-017-6478-y

[43] Predoana L, Calderon-Moreno JM, Anastasescu M, Stoica M, Stanciu I, Preda S, Gartner M, Zaharescu M. Structure and properties of the V-doped TiO_2 thin films obtained by sol-gel and microwave-assisted sol-gel methods. J Sol-Gel SciTechn.2016;78:589-599. DOI: 10.1007/s10971-016-3972-9

[44] Stanciu I, Predoana L, Preda S, Anastasescu M, Calderon-MorenoJM,Stoica M, Gartner M, Zaharescu M. Synthesis method and substrate influence on TiO_2films doped with low vanadium content. Mater SciSemicond Process. 2017;68:118-127. DOI: 10.1016/j.mssp.2017.06.021

[45] Predoana L, Atkinson I, Karaj DA, Odhiambo VO, Bakos LP, TN Kovacs TN, Pandele-Cusu J, Petrescu S, Rusu A, Szilagyi IM, Pokol G, Zaharescu M. Comparative study of the thermal behavior of Sr–Cu–O gels obtained by sol–gel and microwave-assisted sol–gel method. J Therm Anal Calorim. 2020. DOI: 10.1007/s10973-019-09205-5

[46] Mirzaei H, Davoodnia A. Microwave Assisted Sol-Gel Synthesis of MgO Nanoparticles and Their Catalytic Activity in the Synthesis of Hantzsch 1,4-Dihydropyridines. Chin J Catal. 2012;33:1502-1507. DOI: 10.1016/S1872-2067(11)60431-2

[47] Sekularac G, Kosevic M, Drvenica I, Dekanski A, Panic V, Nikolic B. Titanium coated with high-performance nanocrystalline ruthenium oxide synthesized by the microwave-assisted sol–gel procedure.J Solid State Electrochem.2016;20:3115-3123. DOI: 10.1007/s10008-016-3343-z

[48] Dwivedi R, Maurya A, Verma A, Prasad R, BartwalKS.Microwave assisted sol–gel synthesis of tetragonal zirconia nanoparticles. J Alloys Compd.2011;509:6848-685. DOI: 10.1016/j.jallcom.2011.03.138

[49] Hilaire S, Suess MJ, Kranzlin N, Bienkowski K, Solarska R, Augustynski J, Niederberger M. Microwave-assisted nonaqueous synthesis of WO_3 nanoparticles for crystallographically oriented photoanodes for water splitting. J Mater Chem A. 2014;2:20530-20537. DOI: 10.1039/c4ta04793a

[50] Santiago A, Gonzalez A, Iruin JJ, Fernandez-Berridi MJ, Irusta L. Preparation of superhydrophobic silica nanoparticles by microwave assisted sol–gel process. J Sol-Gel Sci Technol. 2012;61:8-13. DOI: 10.1007/s10971-011-2583-8

[51] Colmenares JC, Aramendia MA, Marinas A, Marinas JM, Urbano FJ. Titanianano-photocatalysts synthesized by ultrasound and microwave methodologies: Application in depuration of water from 3-chloropyridine. J MolCatal A: Chem. 2010;331:58-63. DOI: 10.1016/j.molcata.2010.07.014

[52] Alsharaeh EA, Bora T, Soliman A, Faheem Ahmed F, Bharath G, Ghoniem MG, Abu-Salah KM, Dutta J. Sol-Gel-Assisted Microwave-Derived Synthesis of Anatase Ag/TiO_2/GO Nanohybrids toward Efficient Visible Light Phenol Degradation. Catalysts. 2017;7:133. DOI:10.3390/catal7050133

[53] Hernandez R, Duron-Torres SM, Esquivel K, Guzman C. Microwave Assisted Sol-Gel Synthesis and Characterization of $M–TiO_2$ (M = Pt, Au) Photocatalysts. In: Perez Campos R, Contreras Cuevas A, Esparza Munoz R, editors.

Characterization of Metals and Alloys. Cham: Springer; 2017. p.183-189, DOI:10.1007/978-3-319-31694-9_15

[54] Maragatha J, Rajendran S, Endo T, Karuppuchamy S. Microwavesynthesis of metal doped TiO_2 for photocatalyticapplications, J MaterSci: Mater Electron. 2017;28:5281-5287. DOI: 10.1007/s10854-016-6185-7

[55] Hoang LH, Hai PV, Hanh PV, Hai NH, Chen XB, Yang IS. Microwave-assistedsynthesisandcharacterization of $Ti_{1-x}V_xO_2$ (x=0.0-0.10) nanopowders. MaterLett. 2011;65:3047-3050.DOI: 10.1016/j.matlet.2011.06.034

[56] Buch VR, Chawla AK, Rawal SK. Review on electrochromic property for WO_3 thin films using different deposition techniques. Mater Today Proc. 2016;3:1429-1437.DOI: 10.1016/j.matpr.2016.04.025

[57] Dong C, Zhao R, Yao L, Ran Y, Zhang X, Wang Y. A review on WO_3 based gas sensors: Morphology control and enhanced sensing properties. J Alloys Compd. 2020;820:153194. DOI: 10.1016/j.jallcom.2019.153194

[58] Dong P, Hou G, Xi X, Shao R, Dong F. WO_3-based photocatalysts: morphology control, activity enhancement and multifunctional applications. Environ Sci Nano. 2017;4:539-557. DOI: 10.1039/c6en00478d

[59] Hench LL, West JK. The Sol-Gel Process. Chem Rev. 1990;90:33-72. DOI: 10.1021/cr00099a003

[60] Kharade RR, PatilKR, Patil PS, Bhosale PN. Novel microwave assisted sol gel synthesis (MW-SGS) and electrochromic performance of petal like h-WO_3 thin films. Mater Res Bull. 2012;47:1787-1793. DOI: 10.1016/j.materresbull.2012.03.025

[61] Kharade RR, Mali SS, Patil PS, Patil KR, Gang MG, Patil PS, Kim JH, Bhosale PN. Enhanced electrochromic coloration in Ag nanoparticle decorated WO_3 thin films. ElectrochimActa. 2013;102:358-368. DOI: 10.1016/j.electacta.2013.03.123

[62] Kharade RR, Mali SS, Mohite SS, Kondalkar VV, Patil PS, Bhosale PN. Hybrid Physicochemical Synthesis and Electrochromic Performance of WO_3/MoO_3 Thin Films. Electroanalysis. 2014;26:2388-2397. DOI: 10.1002/elan.201400239

[63] Sivakumar R, Raj AME, Subramanian B, Jayachandran M, Trivedi DC, Sanjeeviraja C. Preparation and characterization of spray deposited n-type WO_3 thin films for electrochromic devices. Mater Res Bull. 2004;39:1479-1489. DOI: 10.1016/j.materresbull.2004.04.023

[64] Breedon M, Spizzirri P, Taylor M, Du Plessis J, McCulloch D, Zhu J, Yu L, Hu Z, Rix C, Wlodarski W, Kalantar-zadeh K. Synthesis of nanostructured tungsten oxide thin films: A simple, controllable, inexpensive, aqueous sol-gel method. Cryst Growth Des. 2010;10:430-439. DOI: 10.1021/cg9010295

[65] Ghasemi L, Jafari H. Morphological characterization of tungsten trioxide nanopowders synthesized by sol-gel modified Pechini's method. Mater Res. 2017;20:1713-1721. DOI: 10.1590/1980-5373-MR-2017-0467

[66] Le Houx N, Pourroy G, Camerel F, Comet M, Spitzer D. WO_3 nanoparticles in the 5-30 nm range by solvothermal synthesis under microwave or resistive heating. J PhysChem C. 2010;114:155-161. DOI: 10.1021/jp908669u

[67] Nagy D, Firkala T, Drotar E, Szegedi A, Laszlo K, Szilagyi IM. Photocatalytic WO_3/TiO_2 nanowires:

WO_3 polymorphs influencing the atomic layer deposition of TiO_2. RSC Adv. 2016;6:95369-95377. DOI: 10.1039/c6ra18899k

[68] Firkala T, Keri O, Gaber F, Kocs L, Horvolgyi Z, Nagy D, Zaharescu M, Szilagyi I. Photocatalytic Properties of Hexagonal WO_3 Nanowires Decorated with Gold Nanoparticles. Rev RoumChim. 2017;62:767-773

[69] Nie X, Wei SH, Zhang SB. First-principles study of transparent p-type conductive $SrCu_2O_2$and related compounds. Phys Rev B - Condens Matter Mater. Phys.2002;6:5 1-8. DOI: 10.1103/PhysRevB.65.075111

[70] Banerjee AN, Chattopadhyay KK. Electro-optical properties of all-oxide p-$CuAlO_2$/n-ZnO: Al transparent heterojunction thin film diode fabricated on glass substrate. Cent Eur J Phys. 2008;6:57-63. DOI: 10.2478/s11534-008-0005-5

[71] Kudo A, Yanagi H, Hosono H, Kawazoe H. $SrCu_2O_2$: A p-type conductive oxide with wide band gap. Appl Phys Lett. 1998;73:220-222. DOI: 10.1063/1.121761

[72] Tambunan OT, Tukiman H, Parwanta KJ, Jeong DW, Jung CU, Rhee SJ, Liu C. Structural and optical properties of $SrCu_2O_2$ films deposited on sapphire substrates by pulsed laser deposition. Superlattices Microstruct. 2012;52:774-781. DOI: 10.1016/j.spmi.2012.06.012

[73] Liu JW, Lee SC, Yang CH. Properties of strontium copper oxide films prepared by radio frequency reactive magnetron sputtering with different oxygen partial pressures. Mater Trans. 2007;48:2743-2746. DOI: 10.2320/matertrans.MER2007108

[74] Bobeico E, Varsano F, Minarini C, Roca F. 2003 P-type strontium-copper mixed oxide deposited by e-beam evaporation. Thin Solid Films. 2003;444:70-74. DOI: 10.1016/S0040-6090(03)01023-X

[75] Adachi Y, Takahashi K. MBD Preparation of $SrCuO_2$ Infinite Layer Thin Films. Key Eng Mater. 1999;169-170:163-166. DOI: 10.4028/www.scientific.net/KEM.169-170.163

[76] Deschanvres JL, Millon C, Jimenez C, Khan A, Roussel H, Servet B, Durand O, Modreanu M. Study of the growth and annealing conditions of $SrCu_2O_2$ (SCO) thin films deposited by injection MOCVD. Phys Status Solidi Appl Mater Sci. 2008;205: 2013-2017. DOI: 10.1002/pssa.200778920

[77] Roy B, Perkins JD, Kaydanova T, Young DL, Taylor M, Miedaner A, Curtis C, Kleebe HJ, Readey DW, Ginley DS. Preparation and characterization of sol-gel derived copper-strontium-oxide thin films. Thin Solid Films. 2008;516:4093-4101. DOI: 10.1016/j.tsf.2007.10.002

[78] Ginley DS, Roy B, Ode A, Warmsingh C, Yoshida Y, Parilla P, Teplin C, Kaydanova T, Miedaner A, Curtis C, Martinson A, Coutts T, Readey D, Hosono H, Perkins J. Non-vacuum and PLD growth of next generation TCO materials. Thin Solid Films. 2003;445:193-198. DOI: 10.1016/j.tsf.2003.08.008

[79] Zaharescu M, Mocioiu OC. Infrared Spectroscopy. In: Schneller T, Waser R, Kosec M, Payne D, editors. Chemical solution deposition of functional oxide thin films. Springer, Verlag-Wien; 2013. p. 213-230. DOI: 10.1007/978-3-211-99311-8

[80] Coate J. Interpretation of infrared spectra, A practical approach. In: Meyers RA, editor. Encyclopedia

of analytical chemistry. Wiley, London; 2006. p. 10815-10837. Doi: 10.1002/9780470027318.a5606

[81] Nakamoto K. Infrared and Raman spectra of inorganic and coordination compounds. 4th ed. New York: Wiley; 1986

Chapter 6

Microwave Heating of Low-Temperature Plasma and Its Application

Tetyana Frolova, Vyacheslav Buts, Gennadiy Churyumov, Eugene Odarenko and Vladimir Gerasimov

Abstract

In this chapter, the results of theoretical and experimental studies of the interaction of an electromagnetic field with a plasma (fundamental interaction of the wave-particle type) both in the regime of standing waves (in the case of a resonator) and in the case of traveling waves in a waveguide are presented. The results of computer modeling the distribution of a regular electromagnetic field for various designs of electrodynamic structures are considered. The most attractive designs of electrodynamic structures for practical application are determined. A brief review and analysis of some mechanisms of stochastic plasma heating are given as well as the conditions for the formation of dynamic chaos in such structures are determined. Comparison analysis of microwave plasma heating in a regular electromagnetic field (in a regime with dynamical chaos) with plasma heating by random fields is considered. It is shown, that stochastic heating of plasma is much more efficient in comparison with other mechanisms of plasma heating (including fundamental interaction of the wave-wave type). The results obtained in this work can be used to increase the efficiency of plasma heating as well as to develop promising new sources of electromagnetic radiation in the microwave and optical ranges.

Keywords: Low–temperature plasma, microwave heating, stochastic heating, electromagnetic wave, dynamic chaos, electrodynamic structure

1. Introduction

In recent years, the most promising method for creating low–temperature plasma is the method of microwave heating. As of now, gas–discharge plasma is effectively used in various fields, including microwave electronics, lighting technology, medicine, plasma chemistry, etc. The use of microwave heating to excite gases and to create plasma allows significantly to reduce the overall dimensions of devices. This is a very current task especially under creating small-size devices, including the annular laser gyroscopes, different types of lasers, electrodeless plasma lamps, etc. [1].

Created designs of a miniature helium-neon laser and a source of incoherent optical radiation based on an electrodeless sulfur lamp are shown in **Figure 1** [2, 3]. The unifying factor of these structures is the method of microwave excitation (heating) of the gas mixture in the working volume of the devices. Considering the

a)

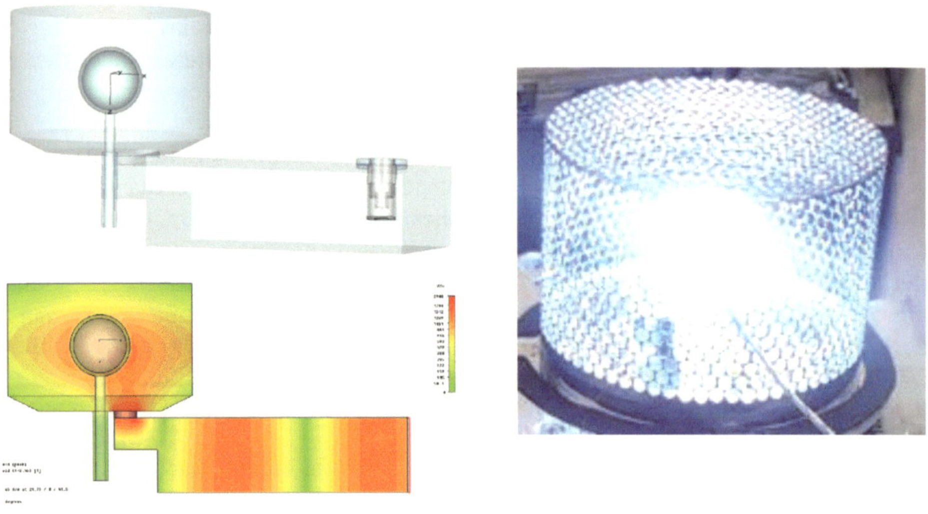

b)

Figure 1.
Examples of structural elements of sources of electromagnetic radiation with microwave excitation. A ring laser gyroscope [2] (a) and an optical radiation source based on an electrodeless sulfur lamp (b) [2].

significant difference in the frequencies of the exciting microwave field, the designs of electrodynamic systems, in the volume of which a regular electromagnetic field is formed, have significant differences. These features of the excitation elements must be taken into account when designing devices, the operation of which is based on the use of gas-discharge media (plasma).

The analysis shows that the main structural elements of these devices are the active medium and the electrodynamic system for the formation of a regular electromagnetic field. A gas or a gas mixture (sometimes with the addition of an impurity, for example, sulfur S8, as in the case of a sulfur lamp) is considered as an active medium with nonlinear properties. Under the action of an external microwave electromagnetic field, the gas mixture is ionized and plasma is formed. At this stage, it is important to understand the ongoing physical processes that underlie the formation of plasma with the necessary quantitative characteristics. To solve such a problem, it is necessary to develop a mathematical model of the plasma, the use of which will make it possible to estimate the quantitative parameters of the gas (gas mixture) to provide the necessary plasma characteristics.

On the other hand, the process of ionization of the active medium (gas or gas mixture) occurs as a result of the action of a regular electromagnetic field. One of the conditions for effective plasma heating is the arrangement of an active medium into an area with the maximum intensity of the electric component of the regular electromagnetic field. To achieve this, it is necessary to know the distribution of the components of the electromagnetic field in the volume of the electrodynamic structure, which requires additional studies of its properties. Currently, there are two approaches to the formation of an electromagnetic field with the aim of its subsequent use for exciting or heating various media: resonance (or a regime of a standing wave) and interference, when two traveling waves add up to give a resultant standing wave [3]. In the first case, a regular electromagnetic field is excited in a resonator. The shape and dimensions of the given resonator are selected from the conditions of the frequency range used, taking into account the maximum intrinsic quality factor of the oscillatory system. The interference method of forming a standing wave involves the use of two traveling waves moving towards each other. To implement this approach, a regular waveguide is used, inside which a bulb with an active medium is placed.

Thus, in order to increase the efficiency of conversion of the energy of the electromagnetic field into the internal energy of the gas–discharge medium, it is necessary to a knowledge of the distribution of a regular electromagnetic field as well as an understanding of the physical processes in the active medium.

In the latter case, it is important to further develop the theory of plasma processes taking into account their chaotization, as well as to understand the conditions for the occurrence of regimes with dynamic chaos (conditions of stochastic heating of plasma).

In the given chapter the theoretical and experimental studies of microwave heating of plasma by an electromagnetic field are discussed. The conditions for increasing the efficiency of the microwave heating at the expense of enhancing the intensity of a regular electromagnetic field or using dynamic chaos mode, including the methods of its achieving (the cases of Cherenkov's and cyclotron resonances) are determined. The computer modeling results of electrodynamic systems are presented and the process of formation of a standing electromagnetic wave in electrodynamic systems (resonator and waveguide) is investigated, conditions for a local increase of intensity of the regular field in the region of the active medium are determined.

2. Computer modeling of the electrodynamic systems

As mentioned above, the energy of the microwave electromagnetic field is used to excite and heat the plasma. Various designs of oscillatory structures (resonators) are used as electrodynamic structures that form fields with the required distribution of power lines, the type and shape of which depend on the frequency of the electromagnetic field used. For example, for the excitation (pumping) of the active medium in helium-neon lasers, electromagnetic oscillations are used, the frequency of which lies in the range of 200 ... 400 MHz with an average microwave power level of ~ 2 ... 5 W. In this case, a classical oscillatory circuit in the form of a flat capacitor is used as an oscillatory electrodynamic system. **Figure 2** shows the design of such a capacitor, the results of simulation and experiment [4]. An analysis of the

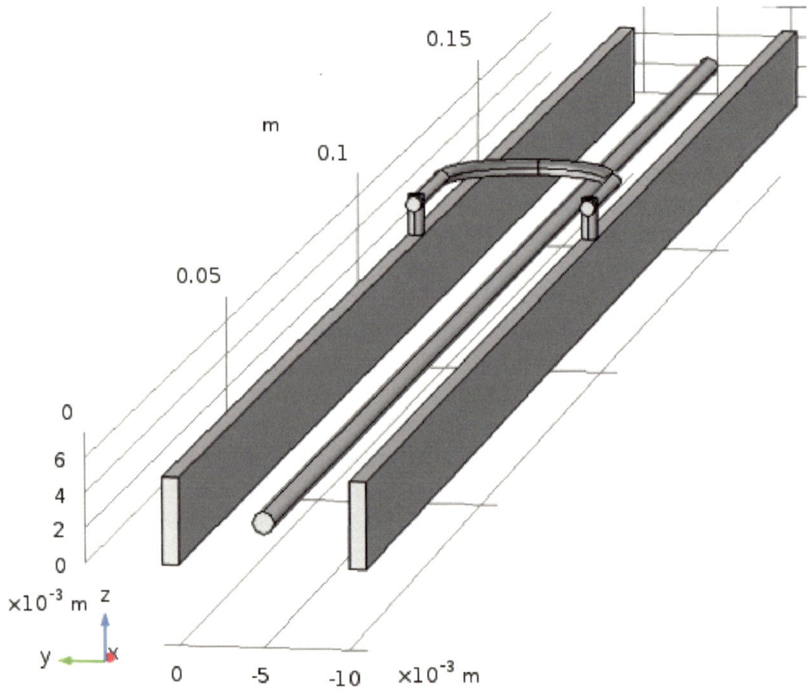

Figure 2.
Capacitor design for exciting a helium-neon laser mixture.

results obtained showed that the pump parameters of the active medium (helium-neon mixture) depend on the dimensions of the flask (its inner diameter and length) containing the active medium and are selected taking into account the maximum efficiency of the pump energy transfer process (see **Figure 1a**).

With the increasing frequency, an electrodynamic system usually modified its shape and dimensions. For frequencies in the range of more than 1000 MHz, cavity resonators are usually used (see **Figure 1b**). In our case, we used a cylindrical resonator having the following geometrical dimensions: a diameter ~ 172 mm and a height ~ 120 mm. As a source of electromagnetic oscillation, there was used a magnetron generator, possessing a frequency of generation ~ 2.45 GHz and the output power ~ 800 W.

Figure 3 illustrates the results of a computer modeling of the two modes of oscillations excited into the cylindrical resonator, namely, the H_{111} and H_{011} modes, correspondingly. The dependences of the reflection coefficient modulus from frequency and the spatial distributions of components of an electric field of the given oscillation modes have been shown.

As is seen, the different modes of oscillations are excited in the cylindrical resonators. As a rule, the given modes have distinct frequencies and possessing different distributions of the electric component of an electromagnetic field. The interest is the oscillations with the frequencies close to the frequency of the magnetron, i.e., to the frequency of 2.45 GHz. On the other hand, it is significant that the selected mode of oscillation had a maximum of the electric field in the area corresponding to the location where must be the bulb with the gaseous mixture. This permits to make an excitation process more effective.

In addition to the resonance excitation method of the plasma for forming a standing wave, we may use the interference of two coherent waves propagating towards each other in a waveguide [3]. Let us consider this approach using an

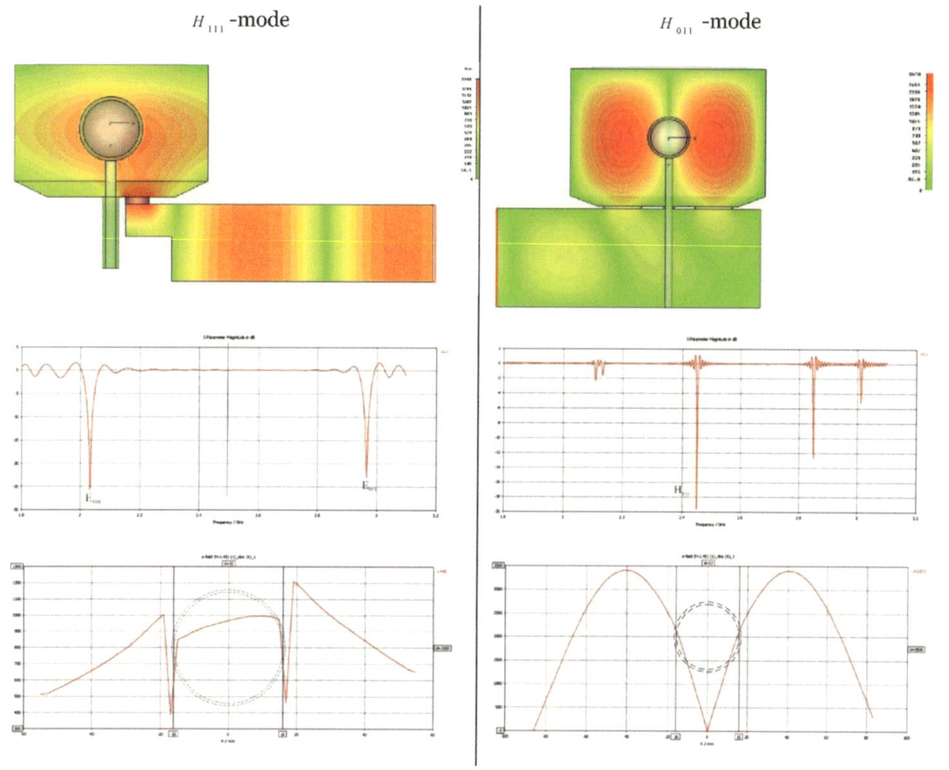

Figure 3.
Results of computer modeling the cylindrical resonator.

example of the waveguide with a cross-section (72x34) mm. This approach may be realized using the waveguide structure that is schematically shown in **Figures 4** and 5. An electromagnetic wave E from the output of the magnetron originates to the input of the waveguide structure at $x = L_0$ (see **Figure** 5). Next at $x = 0$ and $z = 0$, the wave is divided into two waves and that enters the inputs 1 and 2 of a waveguide and propagates towards each other (see **Figures 4** and 5). As a result of the interference of the two coherent waves in the waveguide, the standing wave is formed as it may be seen from **Figure 6**.

For increasing the intensity of the electric component of a total electromagnetic field and enhancing the efficiency of exciting plasma are of interest to view a case of contraction of the narrow size b. The main results of computer modeling the propagation of the waves in the space of the waveguide $L_0 - L_4$ taking into account a change of the size of the narrow wall in the waveguide are shown in **Figure** 7.

As indicated in **Figure** 7, the general regularity of changing the intensity of the total electromagnetic field in the waveguide $E = E_1 + E_2$ is nonlinear and satisfies the condition $E \sim 1/b$, where b – the size of the narrow wall of the waveguide. By varying the high of the narrow wall of the waveguide we may choose a necessary value of the intensity of electric components and thus increasing the efficiency of controlling by a process of plasma heating. Such an approach of increasing the efficiency of the excitation process of the plasma mixture can be used for the choice of an optimum design of the electrodynamic structure. Among possible electrodynamic structures having an enhanced concentration of the electric component of electromagnetic field one can select both the resonant species of such structures and non-resonant. To the first group, we can relate toroid and coaxial resonators. The

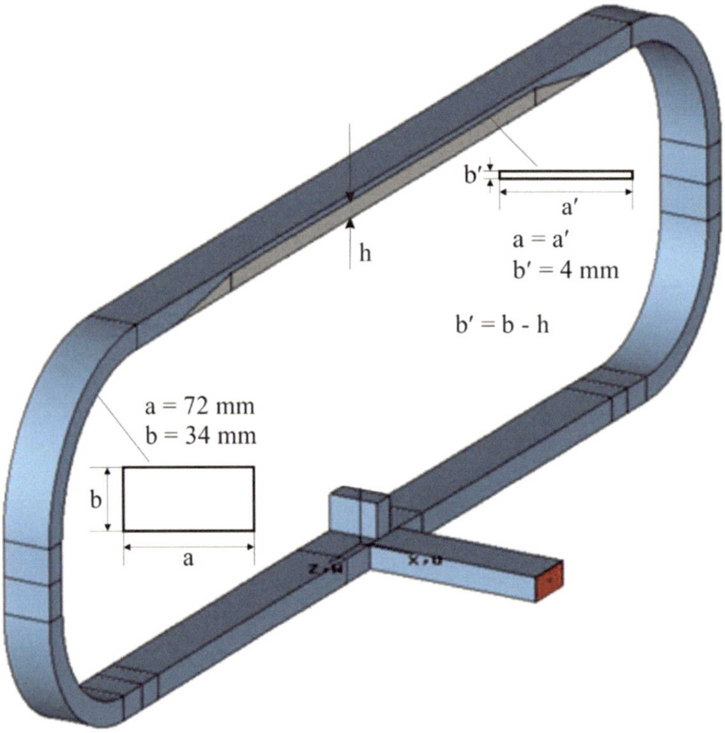

Figure 4.
Schematic image of the waveguide structure with a metal insert having a height h.

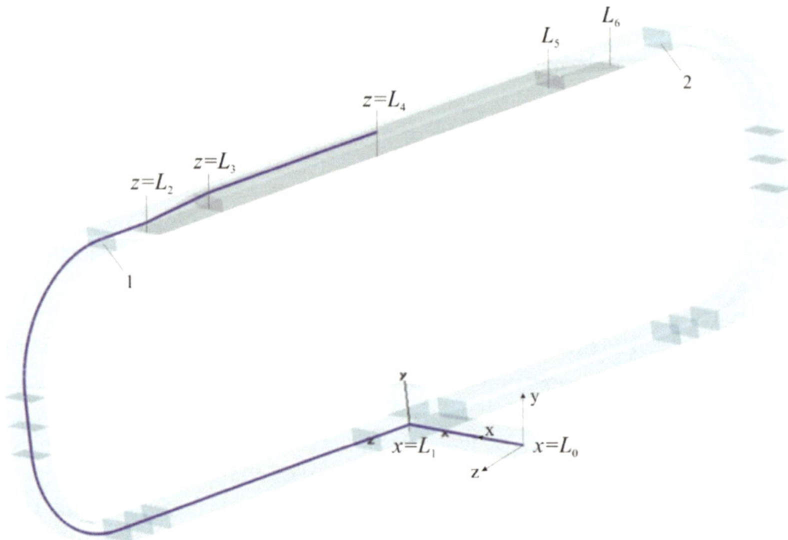

Figure 5.
Image of the curve $L_0 - L_4$ along which the electric field value is calculated.

second group of such structures can be presented by the single- and double-ridged waveguides. Thus, an application of the above-mentioned electrodynamic structures enables improving the process of exciting plasma and enhancing the efficiency of transformation of energy.

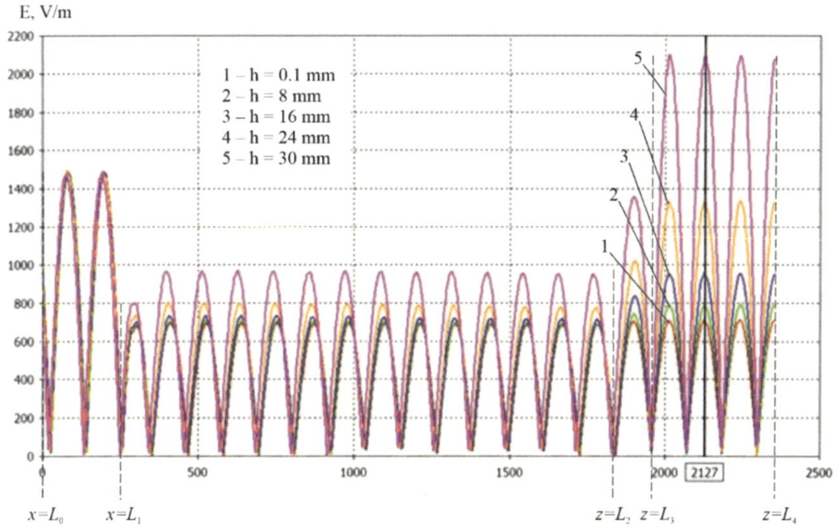

Figure 6.
Distributions of the electric component of the total electromagnetic wave for different values of the height b′ of the waveguide.

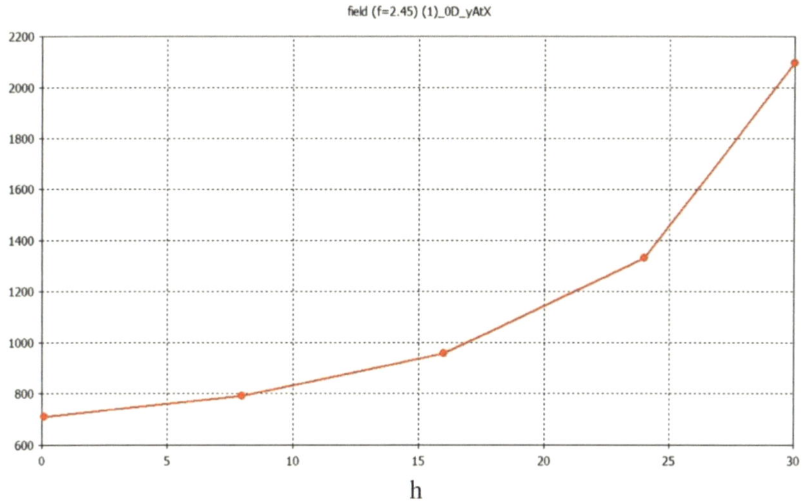

Figure 7.
Dependence the intensity of the electric component from a value of b′ at the point z = 2127 (see ***Figures 4*** *and* ***6****).*

3. Mathematical models and modeling of stochastic processes in plasma

3.1 Mathematical description of the state of the plasma and its model

Due to the variety of processes taking place in a spatially inhomogeneous plasma, an analytical description of a real plasma in the general case is very difficult. Therefore, simplified plasma models are usually considered, stipulating the conditions under which a real plasma can be close to its accepted model.

The state of a real plasma at an arbitrary pressure is determined by a) the concentration of particles of all kinds N (the number of particles per unit volume);

b) their speed distribution functions $N_i(n)$; c) the population of the excited levels N_k (the number of particles per unit volume, excited in the state k); d) the spatial distribution of these quantities.

It is extremely difficult to obtain information about all the listed characteristics since theoretical studies of the state of plasma-like media require the compilation and solution of a system of equations connecting the indicated quantities with external conditions.

The basic equations describing the nonlinear states of plasma have limitations, primarily related to the possibility of obtaining their solutions. Too simple mathematical plasma models also have limited capabilities due to their inability to adequately reflect the behavior of real plasma. The strongest difference between the real state of the plasma and its mathematical description is observed in the so-called boundary zones, where the plasma passes from one physical state to another (for example, from a state with a low degree of ionization to a state with a high degree of ionization). In this case, the plasma cannot be described using simple smooth functions and a probabilistic approach is required to describe it. Effects such as a spontaneous change in the state of plasma are a consequence of the complex nonlinear interaction of charged particles that make up it. Therefore, to describe plasma, models of the state of the plasma are used, which relate the values of its main parameters, and, therefore, determine its basic properties and behavior.

3.1.1 Local thermodynamic equilibrium (LTE) model

To describe the low–temperature plasma in the bulb of an electrodeless sulfur lamp, which is formed under the action of an electromagnetic field, one can use the LTE model. This model makes it possible to qualitatively and quantitatively describe the continuous emission spectrum of the lamp, as well as the distribution of the main physical quantities of sulfuric plasma.

According to the LTE model, the temperature in different elements of the volume of the medium is different, there is a radiation flux outward (the radiation field is anisotropic), but for each element of the volume of the medium, the Boltzmann and Maxwell distributions, as well as the Saha formula, are valid. Moreover, all of them for a given volume include the same local temperature value, which is the same for all types of particles.

Basic equations describing the LTE model:

– the number of atoms or ions in an arbitrary excited state k (population of the state k) is determined by the Boltzmann formula

$$N_k = N_0 \frac{g_k}{g_0} \exp\left(\frac{-E_k}{kT}\right) = N \frac{g_k}{U} \exp\left(\frac{-E_k}{kT}\right), \tag{1}$$

where N_0 – the population of the ground state; g_0 – the statistical weight of this state; g_k – statistical weight of the excited state; E_k – the energy of the excited state, measured from the ground level.

Statistical sums over all energy levels E_n of the corresponding ions (atoms) is equal

$$U = \sum_n g_n \exp\left(-\frac{E_n}{kT}\right). \tag{2}$$

In the case of a single ionization of a gas, the concentrations of atoms, ions, and electrons are related to each other by the Saha formula

$$\frac{N_e N_i}{N_a} = 2\frac{(2p\, m_e)^{\frac{3}{2}}}{h^3}(kT)^{\frac{3}{2}}\frac{U}{U_a(T)}\exp\left(\frac{-E_i}{kT}\right), \tag{3}$$

where m_e – the electron mass; E_i – ionization energy; $U_i(T)$ and $U_a(T)$ are the sums over the states of ions and atoms; g = 2 – the statistical weight of electrons.

In plasma of a complex chemical composition, equation (3) is valid for the ions of each chemical element; chemical reactions can occur there, dissociation and recombination of molecules can occur. All these reactions obey the law of mass action with the same temperature T.

The distribution of particles of any kind i by velocity ν is expressed by the Maxwell function

$$N_i(n) = 4pN_i\left(\frac{M_i}{2pkT}\right)^{\frac{3}{2}}\exp\left(-\frac{M_i n^2}{2kT}\right), \tag{4}$$

where M_i – the mass of particles; $N_i(n)$ – the number of particles (concentration) with velocities ranging from n to n + dn; N_i – concentration equal to

$$N_i = \int_0^\infty N_i(v)\,dv. \tag{5}$$

The pressure p in the plasma is found from the equation of state

$$p = \sum_i N_i kT. \tag{6}$$

Local thermodynamic equilibrium is a state of a plasma in which all distribution functions are in equilibrium, except for one concerning radiation: there is no equilibrium of optical processes, as a result of which the Planck formula turns out to be unsuitable.

LTE is typical for most stationary plasmas obtained under laboratory conditions. Under conditions of LTE plasma, the detailed equilibrium with respect to optical transitions is violated; therefore, it is advisable to consider radiation and absorption separately.

Plasma, in which radiation of a given wavelength is practically not absorbed, is optically thin for this radiation. The radiation intensity J_{ki} of an optically thin plasma in the LTE state within the spectral line with the following frequency n_{ki} may be written as

$$J_{ki} = N_k A_{ki} h n_{ki} = N_0 A_{ki} h n_{ki} \exp\left(-\frac{E_k}{kT}\right). \tag{7}$$

LTE plasma, described by a single parameter T, can exist in a limited pressure range.

The numerical model of the optical radiation source can be a spherical or cylindrical flask made of transparent anhydrous quartz glass filled with a metered amount of sulfur $\sim$ 1 ... 3 mg (and it is also possible to introduce impurities, for example, $CaBr_2$ or indium iodide InI, etc.) and a buffer gas (argon, neon, krypton) under the pressure of $\sim$ 45 ... 170 torrs. By changing the composition of the lamp bulb filling, it is possible to carry out theoretical studies of the output spectral characteristics of optical radiation and their dependence in the optical wavelength range on the microwave pump power, the temperature distribution inside the bulb, plasma electrical conductivity, etc.

Buffer gas (argon) – serves for initial ionization and obtaining a glow discharge (gas pressure is set at the initial stage). We obtain the dependence of the dynamics of changes in pressure in the flask on temperature.

3.2 Stochastic plasma heating

3.2.1 Introduction

By stochastic heating, we mean a process in which, as a result of nonlinear dynamics, plasma particles move chaotically in the fields of regular electromagnetic waves. Their dynamics differ little from the dynamics of particles in random fields. Below, such regimes we will call regimes with dynamic chaos. The conditions for the occurrence of such modes will be the condition of overlapping nonlinear resonances (Chirikov's criterion). It is known that in the vicinity of resonances the dynamics of particles are described by equations of nonlinear oscillators, in particular, by the equation of a mathematical pendulum. Therefore, the algorithm for finding the conditions for the emergence of regimes with dynamic chaos (conditions of stochastic heating) can be described as follows:

1. The conditions for resonant interaction of waves with particles are found. There should be several such resonances.

2. Equations of nonlinear oscillators are determined, which describe the dynamics of particles in the vicinity of resonances.

3. The conditions for the overlap of nonlinear resonances of these oscillators are found.

These conditions will be the conditions of stochastic heating.

Below, this algorithm is used for the case of Cherenkov resonances, as well as for cyclotron resonances.

3.2.2 The case of Cherenkov resonances

3.2.2.1 Heating of particles in the field of several transverse electromagnetic waves

Consider the dynamics of motion of charged particles in the field of several electromagnetic waves. Expressions for the electric and magnetic fields of these waves can be represented in this form

$$\begin{aligned}
\vec{E} &= \sum_n \vec{E}_n, \\
\vec{H} &= \sum_n \vec{H}_n, \\
\vec{E}_n &= \mathrm{Re}\left(\mathcal{E}_n e^{i\psi_n}\right), \\
\vec{H}_n &= \frac{c}{\omega_n}\left[\vec{k}_n \vec{E}_n\right],
\end{aligned} \tag{8}$$

where $\psi_n = \vec{k}_n \vec{r} - \omega_n t$.

These fields satisfy Maxwell's equations. The equations of motion of a charged particle in fields (8) have the traditional form

$$\frac{d\vec{P}}{dt} = e\vec{E} + \frac{e}{c}\left[\vec{v}\vec{H}\right]. \tag{9}$$

It is convenient to write these equations in the following dimensionless variables for both dependent and independent variables: $\omega_n = \omega_n/\omega_0, \dot{\vec{P}} \equiv d\vec{P}/d\tau,\ \tau \equiv \omega_0 t$, $\vec{P} \equiv \vec{P}/mc,\ \dot{\vec{r}} = \vec{v}/c,\ \vec{k}_n \equiv \vec{k}_n c/\omega_0,\ \vec{r} \equiv \omega_0\vec{r}/c,\ \vec{E}_n \equiv e\vec{E}_n/mc\omega_n$ – is the wave force parameter.

Eq. (9) can be conveniently supplemented with the energy equation

$$\dot{\gamma} = \frac{\vec{P}}{\gamma}\frac{e\vec{E}}{mc\omega_0}. \tag{10}$$

Substituting fields (8) into Eqs. (9) and (10) and using these dimensionless variables, we can obtain the following, convenient for further analysis, equations

$$\begin{aligned} \dot{\vec{P}} &= \sum_n E_n\left(\omega_n - \vec{k}_n\dot{\vec{r}}\right) + \sum_n \vec{k}_n\left(\dot{\vec{r}}\,\vec{E}_n\right), \\ \dot{\gamma} &= \frac{\vec{P}}{\gamma}\sum_n \omega_n\vec{E}_n, \end{aligned} \tag{11}$$

where $\vec{E}_n = \mathrm{Re}\left(\vec{\mathcal{E}}_n e^{i\psi_n}\right);\ \psi_n \equiv \vec{k}_n\vec{r} - \omega_n\tau$.

Let us introduce some auxiliary characteristic of the particle, which we will further call the partial energy of the particle, which satisfies the following equation

$$\dot{\gamma}_n = \omega_n\left(\dot{\vec{r}}\,\vec{E}_n\right). \tag{12}$$

From the definition of this partial energy, it follows that it determines the value of the energy that a particle would have if it moved only in the field of one n-th electromagnetic wave. Using the definition of this partial energy, we obtain from Eqs. (11) and (12) the following integral of motion

$$\vec{P} - \sum_n \mathrm{Re}\left(i\vec{\mathcal{E}}_n e^{i\psi_n}\right) - \sum_n \frac{\vec{k}_n}{\omega_n}\gamma_n = \vec{C}. \tag{13}$$

A possible dispersion diagram of three waves interacting with particles is shown in **Figure 8**. This figure shows both the dispersion characteristics of the waves themselves ($\omega_0, \omega_1, \omega_2; k_0, k_1, k_2$) and the dispersion characteristics of the combination waves with which the Cherenkov resonance of plasma particles ($v_{ph1}; v_{ph2}$) occurs. In the general case, Eqs. (11) and (12) together with the integral (13) can be studied only by numerical methods. To obtain analytical results, we will assume that the force parameter of each of the waves acting on the particle is small. In this case, all the characteristics of a particle (its energy, momentum, coordinate, velocity) can be represented as a sum of slowly varying and rapidly changing quantities

$$\begin{aligned} \vec{P} &= \overline{\vec{P}} + \tilde{\vec{P}}, \\ \gamma_n &= \overline{\gamma}_n + \tilde{\gamma}_n. \end{aligned} \tag{14}$$

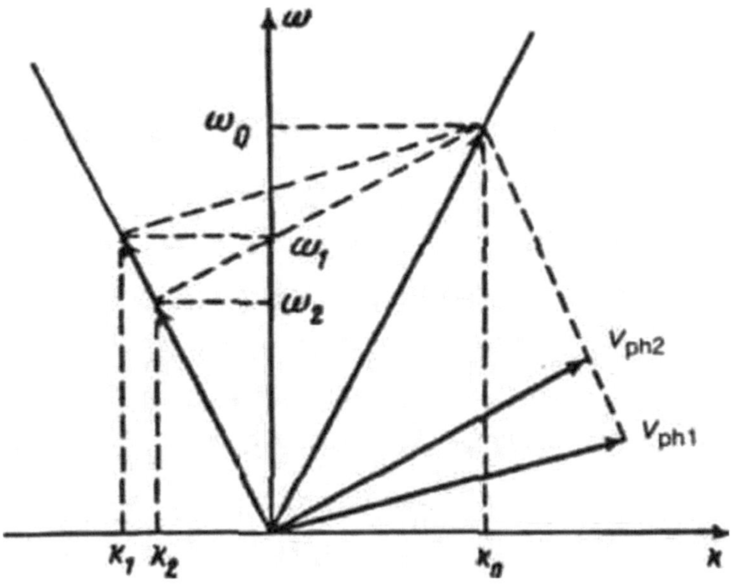

Figure 8.
Dispersion diagram of interacting waves.

In this case, we can get the following expressions and equations that relate fast and slow variables:

$$
\begin{aligned}
\bar{\vec{P}} &= \sum_n \frac{\vec{k}_n}{\omega_n}\bar{\gamma}_n + C,\\
\tilde{\vec{P}} &= \sum_n \mathrm{Re}\left(i\vec{\mathcal{E}}_n e^{i\psi_n}\right) + \sum_n \vec{k}_n \tilde{\gamma}_n/\omega_n,\\
\dot{\tilde{\gamma}}_n &= \omega_n \tilde{\vec{v}\vec{E}}_n = \omega_n \vec{v}\,\mathrm{Re}\left(\vec{\mathcal{E}}_n e^{i\psi_n}\right),\\
\dot{\bar{\gamma}}_n &= \omega_n \overline{\vec{v}\vec{E}}_n,\\
\tilde{\gamma}_n &= \mathrm{Re}\left(\Gamma_n e^{i\psi_n}\right),
\end{aligned}
\tag{15}
$$

where $\Gamma_n = -i\omega_n \overline{\vec{v}\vec{\mathcal{E}}}_n/\dot{\psi}_n$.

The equations for fast variables can be integrated

$$
\begin{aligned}
\tilde{\gamma}_n &= \mathrm{Re}\left[i\omega_n\left(\overline{\vec{v}\vec{\mathcal{E}}_n}\right)e^{i\psi_n}/\omega_n - \vec{k}_n\bar{\vec{v}}\right],\\
\tilde{\vec{P}} &= \sum_n \mathrm{Re}\left\{ie^{i\psi_n}\left[\vec{\mathcal{E}}_n + \vec{k}_n\left(\overline{\vec{v}\vec{\mathcal{E}}_n}\right)/\omega_n\right]\right\}.
\end{aligned}
\tag{16}
$$

The equations for the slow variables take the following form:

$$
\dot{\bar{\vec{P}}} = \sum_{m,n} \vec{k}_n \frac{1}{\gamma}\left[\mathrm{Re}\left(i\vec{\mathcal{E}}_m e^{i\psi_m}\right)\right]\left[\mathrm{Re}\left(\vec{\mathcal{E}}_n e^{i\psi_n}\right)\right] \tag{17}
$$

and

$$\dot{\gamma} = \frac{1}{\gamma}\sum_{m,n} \text{Re}\left(i\vec{\mathcal{E}}_m e^{i\psi_m}\right)\omega_n \text{Re}\left(\vec{\mathcal{E}}_n e^{i\psi_n}\right) =$$
$$= \sum_{m,n}\frac{1}{2\gamma}\omega_n \vec{\mathcal{E}}_n \vec{\mathcal{E}}_m[\cos(\psi_m + \psi_n + \pi/2) + \cos(\psi_m - \psi_n + \pi/2)]. \quad (18)$$

Below, we use the obtained equations and integrals to analyze the dynamics of some physical systems that are of considerable interest.

3.2.2.2 Resonances

In accordance with the algorithm described above, we will find resonances. In addition, we find equations that describe the dynamics of particles in the vicinity of resonances. All waves (1) are transverse and fast. In the original formulation of the problem, there is no mechanism for the resonant interaction of such waves separately with plasma particles. However, plasma particles can have a Cherenkov resonance with a beating wave (with a virtual wave; a combination wave). Indeed, let there be only two fast transverse waves (numbered 1 and 2) among those waves that act on a particle. The beats of these waves form a slow combination wave, the phase velocity of which can be close to the average particle velocity. In this case, the dynamics of particles can be described by the dynamics of a nonlinear pendulum (mathematical pendulum). Let's show it. Let us denote the phase difference of these waves through θ, i.e. $\theta \equiv \psi_1 - \psi_2$. For this phase difference, we obtain the following differential equation

$$\frac{d\theta}{dt} = \vec{\chi}\vec{v} - \Omega = \Delta(\gamma), \quad (19)$$

where $\vec{\chi} \equiv \vec{k}_1 - \vec{k}_2$, $\Omega \equiv \omega_1 - \omega_2$

In this case, we assume that the parameters are close to the conditions of the Cherenkov resonance with the combination wave ($\Omega/\chi \cong v$). The second equation of system (11), taking into account the dynamics of slow and fast variables, can be rewritten as

$$\frac{d\gamma}{d\tau} = \frac{1}{\gamma}\mathcal{E}\cdot\Omega\cdot \cos\theta, \quad (20)$$

where $\mathcal{E} = \vec{\mathcal{E}}_1\vec{\mathcal{E}}_2$.

3.2.2.3 Particle dynamics near resonance

We will assume that the initial energy of a particle exactly corresponds to the Cherenkov resonance of a particle with a combination wave. It means that $\Delta(\gamma_0) = 0$. In addition, we will take into account that as a result of the interaction of waves with particles, the energy of the particle has not changed much. In this case, the resonance detuning $\Delta(\gamma)$ can be expanded into a Taylor series:

$$\Delta(\gamma) = \Delta(\gamma_0) + \delta\gamma\left(\frac{\partial\Delta}{\partial\gamma}\right)_{\gamma_0}. \quad (21)$$

Then Eqs. (19) and (20) will be completely closed and take the following form

$$\frac{d\theta}{d\tau} = \delta\gamma\left(\frac{\partial\Delta}{\partial\gamma}\right)_{\gamma_0}, \qquad \frac{d\delta\gamma}{d\tau} = \frac{\mathcal{E}\Omega}{\gamma_0}\cos\theta. \tag{22}$$

The system of equations (22) is equivalent to the equation of the mathematical pendulum

$$\ddot{\theta} = \left(\frac{\partial\Delta}{\partial\gamma}\right)_{\gamma_0}\frac{\mathcal{E} = \Omega}{\gamma_0}\cos\theta. \tag{23}$$

The half-width of the nonlinear resonance of the pendulum (23) is $\sqrt{(\partial\Delta/\partial\gamma)_{\gamma_0}(\mathcal{E} = \Omega/\gamma_0)}$. If there are many waves (three or more), then each pair of waves can organize a combination wave. The phase velocities of these waves can be easily selected in the required way for efficient particle heating. So if the distance between the phase velocities of the nearest combination waves turns out to be less than the sum of the half-widths of nonlinear resonances, then the dynamics of particles in the field of these waves will be chaotic.

It is enough for us to consider the dynamics of particles in the field of three waves. Two of these waves propagate in the same direction, the third propagates towards them. The condition for overlapping nonlinear resonances can be written as

$$\left(v_{ph_{i+1}} - v_{ph_i}\right) \leq \frac{\sqrt{\mathcal{E}_0}}{\gamma_0^2\sqrt{k_0 v_0}}\left[\sqrt{\mathcal{E}_i\Delta\omega_{0i}} + \sqrt{\mathcal{E}_{i+1}\Delta\omega_{0(i+1)}}\right], \tag{24}$$

where $v_{ph_i} = \Delta\omega_{0i}/(k_0 + k_i),\ i = \{1, 2, ..n\},\ \Delta\omega_{0i} \equiv 1 - \omega_i$.

The left side of inequality (24) describes the distance between nonlinear resonances. The right side represents itself the sum of half–widths of two adjacent nonlinear resonances. If inequality (24) is satisfied, then the dynamics of particles become chaotic. This fact is confirmed by both analytical and numerical studies.

Let us briefly describe the results of a numerical study of the original system of Eq. (11) for the case of interaction of particles with three waves (see **Figure 8**). The dynamics of particles was investigated in a field of small and identical field strengths $\mathcal{E}_i = 0.03$ and with large –$\mathcal{E}_i = 0.3$. In **Figure 9** shows the dependence of the change in energy on time for particles with an initial velocity equal to zero. The wave vectors of the waves were equal: $k_1 = -0.8,\ \ k_2 = -1,\ k_0 = 1.2$. In **Figure 10** shows the temporal dynamics of particle energy with large field strength $\mathcal{E}_i = 0.3$.

Figures 9 and **10** it is seen that at low strengths of the electromagnetic field of the waves, the particle performs regular oscillations, being in one nonlinear Cherenkov resonance with one of the combination waves. With an increase in the field strength under the action of the fields, the particle transitions from resonance to resonance, the dynamics of particle motion is irregular with significant changes in the particle energy.

In this section, it will be shown that using regimes with dynamic chaos, it is possible to propose rather simple and efficient schemes for heating solid-state plasma up to temperatures required for nuclear fusion. Moreover, the heating process proceeds extremely quickly, so that all known plasma instabilities do not have time to develop. To prove the possibility of such heating, we will use all the results obtained above. We will assume that the frequency of the laser radiation that acts on a solid target is much higher than the plasma frequency. Then the results

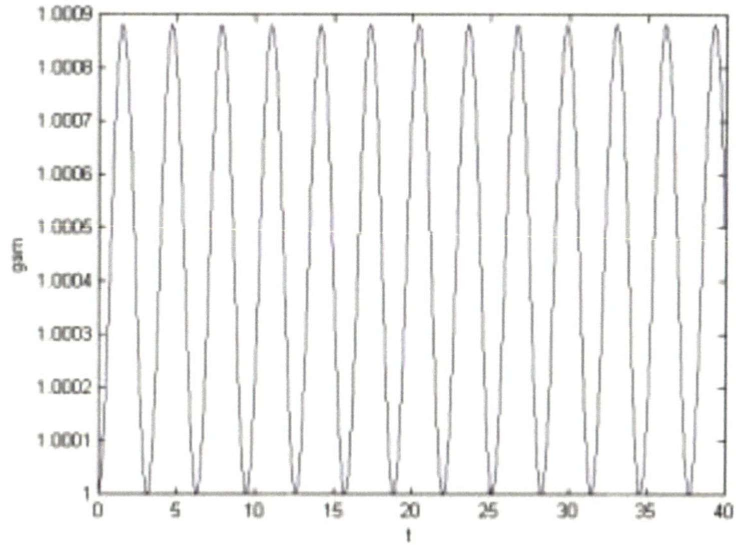

Figure 9.
The energy of one particle at $\mathcal{E}_i = 0.03$ and $k_1 = -0.8, k_2 = -1, k_3 = 1.2$.

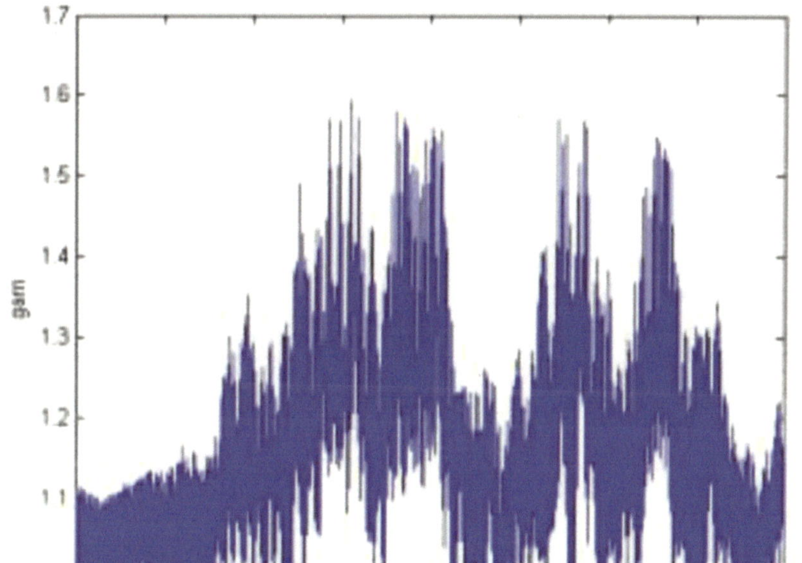

Figure 10.
The energy of one particle at $\mathcal{E}_i = 0.3$ $k_1 = -0.8, k_2 = -1, k_3 = 1.2$.

obtained above can be used in the first approximation. This means that we can assume that condition (24) of overlap of nonlinear Cherenkov resonances is satisfied in the field of laser radiation. If these conditions are met, we can assume that the dynamics of particles is chaotic. Then, by averaging over random phases and random positions of particles, we can find the following expression for the mean square of the change in the energy of particles

$$\left\langle (\Delta\gamma)^2 \right\rangle \approx \mathcal{E}^4 (\Delta\omega)^2 \cdot \tau / \left(4\gamma_0^2\right). \tag{25}$$

Here, the angle brackets denote averaging over phases and positions of particles

$$\langle L\rangle \equiv \frac{1}{2\pi}\int_0^{2\pi} d\left(\vec{k}\vec{r}\right)\cdot \lim \frac{1}{T}\int_{-T}^{T} L\cdot dt, \qquad (26)$$
$$\vec{\mathcal{E}}_n = \vec{E}_n e^{-i\psi_n}.$$

In deriving (25), we assumed that $\Delta\omega_{01} \approx \Delta\omega_{02} \equiv \Delta\omega_0, \mathcal{E}_0 \approx \mathcal{E}_1 \approx \mathcal{E}_2 \equiv \mathcal{E}$ and that the averaging time is much longer than the decoupling time of correlations of particle motion ($\tau >> \tau_k$). The decoupling time of the correlation can be estimated by the value $\tau_k \sim 1/\omega \cdot \ln K$. Here K is the ratio of the width of nonlinear resonances to the distance between them. At $K > 1$, the decoupling time of the correlation is commensurate with the period of the HF field.

A similar analysis of the particle dynamics can be carried out for the case of a large number of waves interacting with particles. The analytical analysis practically does not differ from the one carried out above. Numerical calculations were carried out as well. Let us note the most important results of these studies. The growth rate of the average energy of an ensemble of particles and its maximum energy depends both on the strength of the electromagnetic waves, on the number of combination waves participating in the interaction, as well as on the distance between their nonlinear resonances. Thus, the maximum energy that particles can accumulate in the case of overlap of all Cherenkov resonances from Ncombination waves is determined by the sum of the distances between these resonances

$$\sum_{i=0}^{N-1}\left(v_{ph_{i+1}} - v_{ph_i}\right) = v_{ph_N} - v_{ph_0}. \qquad (27)$$

3.2.2.4 Comparison of heating efficiency

It is of interest to compare the efficiency of plasma heating by fields of regular electromagnetic waves (in a regime with dynamic chaos) with plasma heating by random fields. In random fields, we can write the following equation for the particle energy

$$\frac{d\gamma}{d\tau} = \left(\vec{v}\vec{\mathcal{E}}_n\right). \qquad (28)$$

Here $\vec{\mathcal{E}}_n$ – the field strength of the random wave.

Under the same assumptions under which formula (25) was obtained, we find

$$\left\langle (\Delta\gamma)^2 \right\rangle = v^2 \mathcal{E}_n^2 \tau. \qquad (29)$$

Let us assume that the energy in the field of the noise wave is equal to the energy of the field of coherent radiation. In this case ${\mathcal{E}_n}^2 \cdot \Delta\omega_n = \mathcal{E}^2 \cdot \Delta\omega$. Here $\Delta\omega << \Delta\omega_n$ is the width of the spectrum of the noise field, $\Delta\omega = \omega/Q$ is the width of the spectrum of coherent radiation, $Q$is the Q–factor of the optical resonator ($Q \sim 10^6 - 10^7$).

$$K \equiv \frac{\left\langle (\Delta\gamma)^2 \right\rangle}{\left\langle (\Delta\gamma)^2 \right\rangle_n} > \frac{\mathcal{E}^2 (\Delta\omega_0)^2 Q}{4\gamma_0^2 v_0^2}. \qquad (30)$$

In the vast majority of cases $K >> 1$.

It should be noted that many other heating mechanisms are also less efficient than heating in the dynamic chaos regime. In particular, one can point to the well–known turbulent heating. In turbulent heating schemes, radiation incident on plasma as a result of nonlinear processes excites random fluctuations of fields in the plasma. It is these random fluctuations that heat the plasma particles. As we saw above, this mechanism is less efficient than dynamic heating. In addition, the transformation of regular fields incident on the plasma into random fields requires a significant time.

The closest to the one considered is the scenario of plasma heating, which is associated with collisions of particles of dense (solid-state) plasma. The collision frequency, as is known, is proportional to the plasma density $n = 10^{22} cm^{-3}$ and at a temperature T = 7 keV is $v = 10^{12} s^{-1}$. If the frequency of the laser radiation $\omega = 5 \times 10^{15} s^{-1}$ and the amplitude of the laser wave $\mathcal{E} = 0.1$, then the heating of the plasma to a temperature of 7 keV occurs in a time $\Delta t_H = 2 \times 10^{-14} s$, i.e. in a time significantly shorter than the time of collision between particles. Thus, there is a range of laser radiation and plasma parameters at which dynamic heating is much more efficient than other heating mechanisms.

Let us estimate the possibility of using dynamic heating of solid–state plasma to thermonuclear temperatures. In this case, we need to heat the plasma ions. In this case, direct dynamic heating of ions is ineffective. Indeed, as follows from formula (25), this time is proportional to the fourth power of the mass ($\tau_H \sim (m_i)^4$).

In this case, the ion heating scheme may look as follows: the laser field $\mathcal{E} = 0.1, \omega = 5 \cdot 10^{15} s^{-1}$ heats plasma electrons $n = 10^{22} cm^{-3}$ to a temperature of 7 keV.

This heating takes place over time $t < 10^{-13} s$. During the time $t \sim 10^{-9} s$, the heated electrons transfer their energy to the ions. This time is rather short. During this time, a solid-state target of radius r = 0.1 will not increase its size too much. Note that the rapid heating of electrons and the rapid transfer of energy from electrons to ions make it possible to avoid the development of plasma instabilities.

3.2.3 Plasma heating in an external magnetic field

We saw above that the dynamics of charged particles in the field of a combination wave in the vicinity of the Cherenkov resonance of particles with a combination wave is described by the equation of a mathematical pendulum. If there are several combination waves (we saw above that three transverse electromagnetic waves can generate two combination waves), then to describe the dynamics of particles, it is necessary to analyze a model that contains two equations of a mathematical pendulum. As we saw above, stochastic instability developed when the nonlinear resonances of these mathematical pendulums crossed (see [5, 6]). The particle dynamics became random. Thus, in this model of the interaction of charged particles with electromagnetic waves, the result turned out to be analogous to the motion of particles in a random field. Above, using the example of plasma heating by three laser waves, an expression was obtained that characterizes the efficiency of plasma heating in the field of three regular laser waves. Another common scheme for realizing plasma heating is that the plasma is placed in an external constant magnetic field. To analyze the appearance of conditions for effective plasma heating in such installations based on regimes with dynamic chaos, we note that the presence of an external magnetic field leads to the fact that regimes with dynamic chaos can be realized even when the plasma is exposed to only one external electromagnetic wave It turns out that the role of a large number of waves, in this case, is played by resonances (cyclotron resonances), and also that the dynamics of particles in the vicinity of cyclotron resonances is described by the model of a mathematical pendulum. Overlapping of nonlinear cyclotron resonances leads, as above, to the development of local instability (stochastic instability). As a result, we get method for effective plasma heating [7–11].

Consider a charged particle (electron) that moves in an external constant magnetic field H_0 of magnitude directed along the axis z and in the field of an electromagnetic wave of arbitrary polarization. The components of the electric and magnetic fields of such a wave can be represented as

$$\vec{E} = \mathrm{Re}\left(\vec{\mathrm{E}}_0 e^{i\psi}\right), \quad \vec{H} = \mathrm{Re}\left(\frac{1}{k_0}\left[\vec{k}\vec{E}\right]\right), \tag{31}$$

where $\psi \equiv \omega t - \vec{\mathrm{k}}\vec{\mathrm{r}}$, $\vec{\mathrm{E}}_0 = \vec{\alpha}E_0$; $\vec{\alpha} = \{\alpha_x, i\alpha_y, \alpha_z\}$ – wave polarization vector; $k_0 = \omega/c$; ω, $\vec{k}$ – frequency and wave vector of the wave. We introduce the following dimensionless variables: $\vec{p}_1 = \vec{p}/mc$, $\vec{k}_1 = \vec{k}/k_0$, $\tau = \omega t$, $\vec{r}_1 = k_0\vec{r}$, $\vec{\varepsilon} = e\vec{E}_0/mc\omega$, $\vec{v}_1 = \vec{v}/c$, $v_{ph1} = v_{ph}/c = \omega/kc$.

Without loss of generality, we can assume that the vector $\vec{k}$ has only two nonzero components k_x and k_z. The equations of motion of a particle can be reduced to the form

$$\dot{\vec{P}} = \left(1 - \frac{\vec{k}\vec{p}}{\gamma}\right)\mathrm{Re}\left(\vec{\varepsilon}e^{i\Psi}\right) + \frac{\vec{k}}{\gamma}\mathrm{Re}\left(\vec{p}\vec{\varepsilon}\right)e^{i\psi} + \frac{\omega_H}{\gamma}\left[\vec{p}\vec{e}\right], \quad \dot{\vec{r}} = \vec{p}/\gamma, \quad \dot{\psi} = \vec{k}\vec{p}/\gamma - 1, \tag{32}$$

where $\tau \equiv \omega t$, $\vec{e} \equiv \vec{H}_0/H_0$; $\omega_H \equiv eH_0/mc\omega$; $\psi = \vec{k}\vec{r} - \tau$ [7].

The last term on the right–hand side of the first vector equation describes the Lorentz force that acts on a particle in a constant external field. Multiplying the first of equations (32) by $\vec{p}$ and taking into account that $p^2 = \gamma^2 - 1$, we obtain the following equation for changing the particle energy

$$\dot{\gamma} = \mathrm{Re}\left(v\vec{\varepsilon}\right)e^{i\psi}. \tag{33}$$

Using this equation, from equations (32) we find the following integral of motion

$$\vec{p} - \mathrm{Re}\left(i\vec{\varepsilon}e^{i\psi}\right) + \omega_H\left[\vec{r}\vec{e}\right] - \vec{k}\gamma = const. \tag{34}$$

For what follows, it is convenient to pass to new variables $p_\perp, p_\|, \theta, \xi$ and η, which are related to the old following ratios

$$p_x = p_\perp\cos\theta, \quad p_y = p_\perp\sin\theta, \quad p_z = p_\|, \quad x = \xi - \frac{p_\perp}{\omega_H}\sin\theta, \quad y = \eta + \frac{p_\perp}{\omega_H}\cos\theta. \tag{35}$$

In these variables, Eqs. (32) taking into account the integral (34) take the form

$$\dot{p}_{\perp} = (1 - k_z v_z)\sum_n \left(\varepsilon_x \frac{n}{\mu} J_n - \varepsilon_y J_n'\right)\cos\theta_n + k_x v_z \varepsilon_z \sum_n \frac{n}{\mu} J_n \cos\theta_n,$$

$$\dot{\theta} = -\frac{\omega_H}{\gamma} + \frac{(1 - k_z v_z)}{p_{\perp}} \sum_n \left(\varepsilon_x J_n' - \varepsilon_y \frac{n}{\mu} J_n\right)\sin\theta_n + \frac{k_x v_{\perp}}{p_{\perp}} \varepsilon_y \sum J_n \sin\theta_n + \frac{k_x v_z}{p_{\perp}} \varepsilon_z \sum_n J_n' \cdot \sin\theta_n,$$

$$\dot{p}_{\parallel} = \sum_n \cos\theta_n [\varepsilon_z J_n + (k_z v_{\perp} \varepsilon_x - k_x v_{\perp} \varepsilon_z)] \frac{n}{\mu} J_n - k_z v_{\perp} \varepsilon_y J_n',$$

$$\dot{\gamma} = \sum_n \cos\theta_n \left[J_n \left(v_{\perp} \varepsilon_x \frac{n}{\mu} + v_z \varepsilon_z\right) - v_{\perp} \varepsilon_y J_n'\right],$$

$$\dot{z} = v_z,$$

$$\theta_n = k_z z + k_x \xi - n\theta - \tau. \tag{36}$$

In obtaining (36), we used the expansion

$$\cos(x - \mu \sin\theta) = \sum_{n=-\infty}^{\infty} J_n(\mu) \cos(x - n\theta), \tag{37}$$

where $\mu = k_x p_{\perp} / \omega_H$.

At ($\varepsilon_0 < < 1$) the effective interaction of the particle with the wave occurs when one of the resonance conditions is satisfied

$$\Delta_s(\gamma_o) \equiv k_z v_{z0} + s \frac{\omega_H}{\gamma_0} - 1 = 0. \tag{38}$$

Assuming condition (38) had performed and introducing the resonant phase $\theta_s = s\theta - \tau$ from the system of equations (36), after averaging, we obtain the following equations of motion

$$\begin{aligned} \dot{p}_{\perp} &= \frac{1}{p_{\perp}} (1 - k_z v_z) W_s \cdot \varepsilon_0 \cos\theta_s, \\ \dot{p}_z &= \frac{1}{\gamma} k_z W_s \varepsilon_0 \cos\theta_s, \\ \dot{\theta}_s &= \Delta_s \equiv k_z v_z + s \frac{\omega_H}{\gamma} - 1, \\ \dot{\gamma} &= \frac{\varepsilon_0}{\gamma} W_s \cdot \cos\theta_s, \end{aligned} \tag{39}$$

where $W_s \equiv \alpha_x p_{\perp} \frac{s}{\mu} J_s - \alpha_y p_{\perp} J_s{}' + \alpha_z p_z J_s$.

3.2.4 The condition for the appearance of dynamic chaos (the condition of stochastic heating)

We will assume that the particle energy changes little as a result of the interaction with the electromagnetic wave ($\gamma = \gamma_0 + \tilde{\gamma}$), $\tilde{\gamma} < < \gamma_0$, and the resonance condition (38) is exactly satisfied for a particle with energy γ_o. Then, doing decomposition $\Delta_s(\gamma)$ near γ_o the last two equations of the system (39), we obtain a closed system of two equations for determining $\tilde{\gamma}$ and θ_s

$$\frac{d\tilde{\gamma}}{d\tau} = \frac{\varepsilon_0}{\gamma_0} W_s \cos\theta_s,$$
$$\frac{d\theta_s}{d\tau} = \frac{k_z^2 - 1}{\gamma_0} \tilde{\gamma}. \tag{40}$$

Equations (40) represent the equation of a mathematical pendulum. Of them, we find the width of the nonlinear resonance

$$\Delta\dot{\theta}_s = 4\sqrt{\varepsilon_0\left(k_z^2 - 1\right) \cdot W_s/\gamma_0^2},$$
$$\Delta\tilde{\gamma}_s = 4\sqrt{\varepsilon_0 W_s/\left(k_z^2 - 1\right)}. \tag{41}$$

To find the distance between resonances, we write the resonance conditions (38) and the averaged conservation law (34) for two adjacent resonances (see [6, 7, 11])

$$k_z p_{s+1} + (s+1)\omega_H - \gamma_{s+1} = 0,$$
$$\gamma_{s+1} - p_{s+1}/k_z = C,$$
$$k_z p_s + s\omega_H - \gamma_s = 0,$$
$$\gamma_s - p_s/k_z = C. \tag{42}$$

From these conditions, we find the following value of the distance between resonances

$$\delta\gamma = \omega_H/\left(1 - k_z^2\right). \tag{43}$$

From expressions (42) and (43) it follows that in carrying out the inequality

$$\varepsilon_0 > \omega_H^2/4\left[\sqrt{W_s} + \sqrt{W_{s+1}}\right]^2\left(1 - k_z^2\right), \tag{44}$$

nonlinear resonances overlap. A regime of stochastic instability sets in, and inequality (44) is the condition for stochastic heating of plasma particles. Note that the width of the nonlinear resonance as well as the distance between resonances must be calculated along with the integrals of motion (see **Figure 11**). In this figure, the dotted lines show the boundaries of nonlinear resonances (the position of the separatrices), the solid lines are the cyclotron resonances themselves, and the bold arrow denotes the integral. In all cases, the particle dynamics run according to integrals. When the integral line coincides with the resonance line, the autoresonance condition occurs. Under autoresonance conditions, particles can resonantly acquire unlimited energy.

3.2.5 Experimental studies of stochastic heating of plasma in a constant magnetic field

After theoretical work, a large series of experimental studies of stochastic plasma heating was carried out. At the same time, theoretical estimates showed that for stochastic heating of the plasma by the field of one external electromagnetic wave, the field strength of this wave should be sufficiently high ($\sim 10^6$ V/cm) [8]. This result, for example, follows from an analysis of conditions (44). Additional numerical studies have shown that if several waves are excited in the experimental setup, for example, two waves with the same frequencies but different wave vectors, then

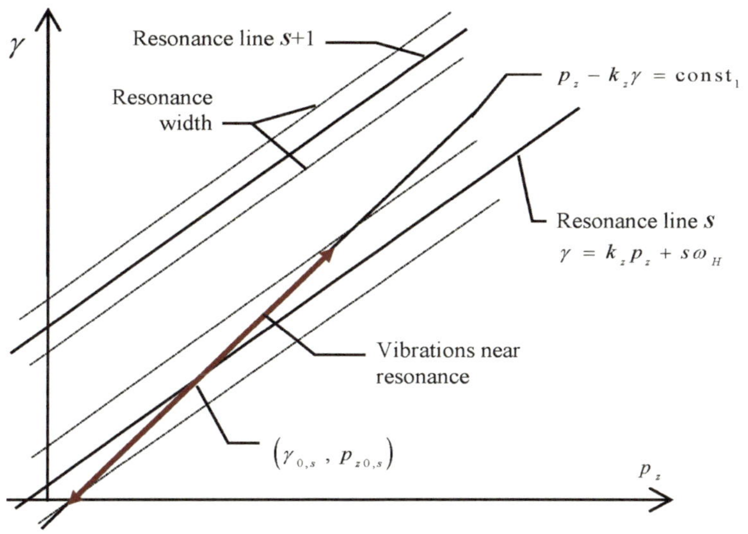

Figure 11.
Location of resonances and one of the integrals (27) on the plane (p_z, γ).

for the occurrence of conditions for stochastic heating, the field strengths significantly decrease ($\sim 10^4$ V/cm) [10, 11].

The main experiments were carried out on a setup, the scheme of which is described in detail in [9]. The main element of this setup is a cylindrical resonator, the general view of which is shown in **Figure 12**. The resonator is made of a copper tube with an inner diameter of 16 cm and a length of 66 cm. The modes H_{10x} and H_{20x} are excited in this resonator. The central axis of the resonator coincided with the direction of the external inhomogeneous magnetic field. This field formed a magnetic trap. The mirror ratio was chosen equal to 1.2 ... 2. The length of the uniform part of the magnetic field of the trap in the cavity was varied from 25 to 66 cm. A loop probe was located in the central part of the cavity. Plasma in the cavity was generated by an electron beam with an energy of 400–600 eV and a current of 60–100 mA due to a beam–plasma discharge [9]. The pressure in the resonator could be regulated in the range of 10^{-4} ... 10^{-6} mm Hg. In the main series of experiments, the plasma density was within $\sim 10^9$ cm^{-3}.

The sequence of working of the equipment in time is as follows. An electron beam is injected into the cavity. As a result of the beam–plasma discharge, a plasma is formed with a density of up to $\sim 10^{11}$ cm^{-3}. The resonator was excited at a frequency of $\sim$ 2.7 GHz, for which the cyclotron resonance was performed at a magnetic induction at the minimum of the trap equal to $\sim$ 0.1 T. The length of the uniform part of the magnetic field of the trap in the resonator was varied from 25 to 66 cm. The oscillation power of the magnetron could be varied in the range 0.1–1.0 MW in a pulse with a duration of 1.8 μs. and was fed into the resonator through a waveguide with a cross-section of 72x34 mm. By varying the delay time between the electron beam pulse and the high-frequency power pulse, the required plasma density was selected in the range $10^7 - 10^9$ cm–3 at a pressure of $10^{-5} - 10^{-4}$ mm Hg. Argon was used as a plasma-forming gas.

The experiments investigated the fluxes of microwave, optical and X–ray radiation. Simultaneously, using a set of foil plates (up to 15 layers of aluminum foil), electron fluxes with energies up to 1 MeV were recorded. The results are shown in **Figure 13**. Estimation of the electron energy at the maximums of the X–ray radiation intensity showed that at t = 2 μsec the electron energy reaches 100–150 keV, while at t = 1 μsec the electron energy is 8–10 times higher ($\sim$ 1 MeV).

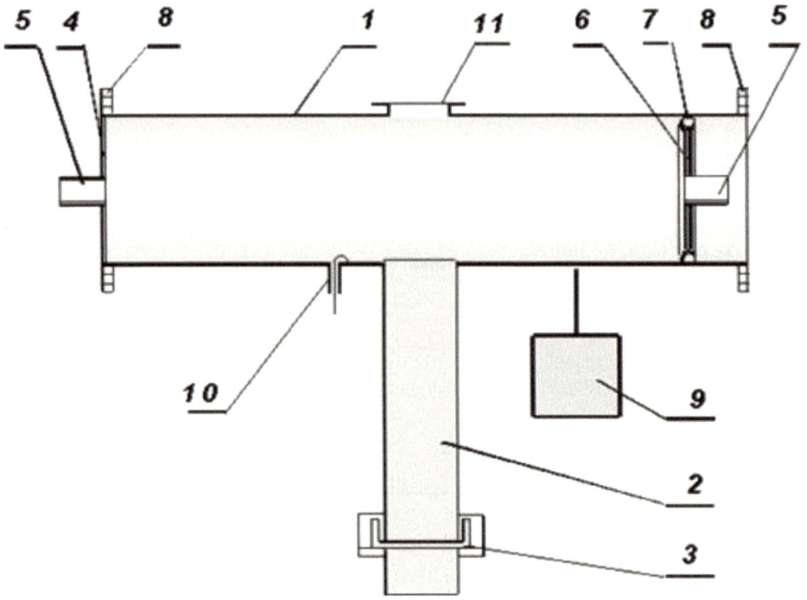

Figure 12.
Some elements of the resonator: 5 – below-cutoff waveguide; 6 – movable piston; 9 – gas supply system; 10 – loop sealed microwave probe; 11 – vacuum window made of Lavsan film with mesh.

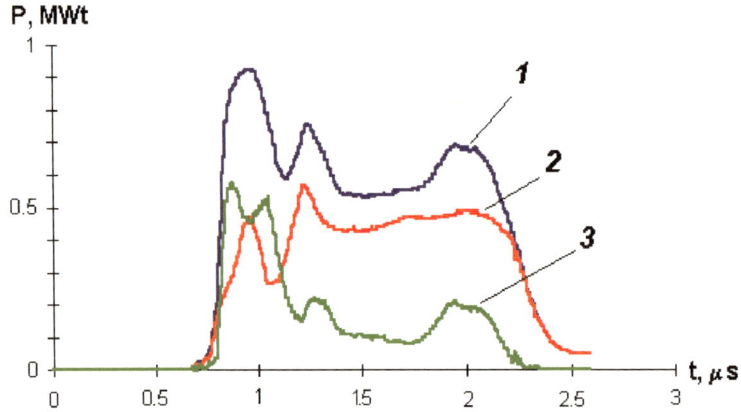

Figure 13.
High-frequency power oscillogram: 1 – incident wave; 2 – backward wave; 3 – absorbed wave.

4. Conclusions

In this chapter, there were considered different approaches to exciting plasma by a regular electromagnetic field. As a microwave source, there was used a magnetron generator as well as two types of electrodynamic structures: resonators (the cylindrical resonator) and waveguides (the rectangular waveguide). An application of the given electrodynamic structures allowed the formation of an electromagnetic field needed for effective exciting plasma in the area of location of an active medium (for example, a bulb with gases mixture). The carried out investigations have pointed to distinct aspects of forming a regular electromagnetic field and its excitation as well as the features of plasma heating. It is significant that for exciting

DOI: http://dx.doi.org/10.5772/intechopen.97167

plasma and enhancing efficiency in its heating it is necessary to optimize not only the shape and special distribution of the electric field in the cavity but also its intensity. In this regard, a preference is given to the resonant electrodynamic structures having a concentrated capacity as a parameter. Among such structures is the coaxial resonator loaded on a capacity as well as the toroidal resonator. In the case of an application of the interference method for forming an electromagnetic field and enhancing the effectiveness of plasma heating a great interest is using the single- and double-ridged waveguides instead of the regular waveguides. On the other hand, a comparison of the theoretical and experimental data showed that the most effective is heating of any plasma when the interaction of plasma particles with regular electromagnetic waves occurs in the dynamic chaos regime. Note that the described mechanisms relate to the interaction of the wave-particle type. Another fundamental interaction (the interaction of wave-wave type) can also be used to heat the plasma. But in that event, such a heating mechanism (turbulent heating) contains two stages. At the first stage, the energy of regular waves is transformed into the energy of less efficient noise vibrations.

The experimental implementation of the conditions for stochastic heating of plasma by the field of regular electromagnetic waves with a high rate of energy transfer from electromagnetic waves to the energy of thermal motion of plasma electrons has been carried out. It is shown that the average energy of plasma electrons reached values $\sim$ 1 MeV in times less than 1 μsec.

Also, it is necessary to note that in the experiment stochastic heating and, accordingly, X-ray radiation from the plasma was observed only when several spatial modes were excited in the resonator or when the resonator was excited by two frequencies. This fact is in full agreement with the results of the analysis of theoretical models.

Author details

Tetyana Frolova[1*], Vyacheslav Buts[2,3], Gennadiy Churyumov[1,4], Eugene Odarenko[1] and Vladimir Gerasimov[1]

1 Kharkiv National University of Radio Electronics, Kharkiv, Ukraine

2 National Science Center 'Kharkiv Institute of Physics and Technology', Kharkiv, Ukraine

3 Radio Astronomy Institute, National Academy of Sciences of Ukraine, Kharkiv, Ukraine

4 Harbin Institute of Technology, Harbin, China

*Address all correspondence to: tetyana.frolova@nure.ua

References

[1] J. E. Harry, Introduction to Plasma Technology: Science, Engineering, and Applications. Wiley, 2013. 232 p. ISBN: 978-3-527-32763-8

[2] Ring laser gyroscope [Internet]. 2020. Available from: https://en.wikipedia.org/wiki/Ring_laser_gyroscope [Accessed: 2020–09–28]

[3] Churyumov Gennadiy and Tetyana Frolova, "Chapter 5. Microwave Energy and Light Energy Transformation: Methods, Schemes and Designs," in Emerging Microwave Technologies in Industrial, Agricultural, Medical and Food Processing / Gennadiy Churyumov, Tetyana Frolova // Book edited by: Kok Yeow You, Ed. Rijeka: InTech, 2018, pp. 75-91. DOI: 10.5772/intechopen.73755

[4] Frolova Tetyana, HF excitation of plasma sources of optical radiation / Tetyana Frolova, Kostiantyn Vasko // Applied Radio Electronics: Sci. Journ. – 2019. – Vol. 18, № 1, 2. – P. 66–72.

[5] V.A. Buts, K.N. Stepanov. Stochastic plasma heating by laser light. JETP Lett. Vol. 58, N 7, 10 Okt. 1993. 524–527.

[6] V.A. Buts. Overview. Problems of Theoretical Physics. Series. Problems of theoretical and mathematical physics. Regular and chaotic dynamics of charged particles in wave-particle interactions. V. 2. Kharkiv, 2017, p. 122-241.

[7] V.A. Balakirev, V.A. Buts, A.P. Tolstoluzhskii, Yu. A. Turkin. Charged-particle dynamics in the field of two electromagnetic waves. JETP, Vol. 68 (4), April 1989, 710–717.

[8] A.N. Antonov, V.A. Buts, O.F. Kovpik, E.A. Kornilov, O.V. Manuilenko, V.G. Svichenskii, K.N. Stepanov. Stochastic heating of plasma at electron cyclotron resonance. JETP LETTERS, v. 69, N 11, 10 june 1999, p. 851–857.

[9] A.N. Antonov, V.A. Buts, I.K. Kovalchuk, O.F. Kovpik, E.A. Kornilov, V.G. Svichensky, D.V. Tarasov. Regular and stochastic decays of waves in a plasma resonator. Plasma Physics, 2012, vol. 38, T 8, p. 693–708

[10] A.N. Antonov, V.A. Buts, I.K. Kovalchuk, O.F. Kovpik, E.A. Kornilov, V.G. Svichensky, D.V. Tarasov. The dynamics of decay processes in rare plasma. Problems of atomic science and technology, 2009. N 1. Series: "Plasma Physics"(15). p.104–106.

[11] V.A. Buts, V.V. Kuzmin, A.P. Tolstoluzhsky. Features of the dynamics of particles and fields at cyclotron resonances. ZhETF, 2017, volume 152, no. 4 (10), p. 767-780.

Chapter 7

Experimental Investigation on the Effect of Microwave Heating on Rock Cracking and Their Mechanical Properties

Gaoming Lu and Jianjun Zhou

Abstract

Due to various advantages including high efficiency, energy-saving, and having no secondary pollution (no dust or noise), the technology of microwave-induced fracturing of hard rock has been considered as a potential method for rock fracturing and breaking. Realizing microwave-assisted mechanical rock cutting using the microwave-induced hard rock fracturing technique can prolong the mechanical life and improve the efficiency of rock-breaking operations. For example, to realize microwave-assisted TBM excavation for hard rock tunnel. At present, this technology is still in the laboratory research stage. By summarizing the research results of relevant scholars in this field, this paper generalizes the mechanism of microwave heating of rock, microwave heating system, heating characteristics, and the effect of microwave heating on rock cracking and mechanical properties. Microwave heating causes microscopic cracks on the surface of the rock and microscopic cracks inside the rock. The higher the microwave power, the longer the irradiation time, the more serious the cracks propagation. Uniaxial compressive, Brazilian tensile, and point load strengths all decreased with increasing microwave irradiation time at rates that were positively related to the power level. The conventional triaxial compressive strength of basalt samples decreased linearly with microwave irradiation time, and the higher the confining pressure, the smaller the reduction in the strength of basalt samples after microwave treatment. In addition, the elastic modulus and Poisson's ratio of basalts decreased in a quasi-linear manner with the growth of microwave irradiation time under uniaxial compression. While microwave irradiation has a slight influence on elastic modulus and Poisson's ratio under triaxial compression. The cohesion decreases with increasing microwave irradiation time and shows an approximately linear decrease over time.

Keywords: microwave heating, TBM excavation, temperature characteristics, crack propagation, mechanical properties

1. Introduction

New, and high-efficiency technology for rock breaking is required in the field of tunnel excavation, mining and mineral processing. The technology of microwave-induced fracturing of hard rock has been considered as a potential method for

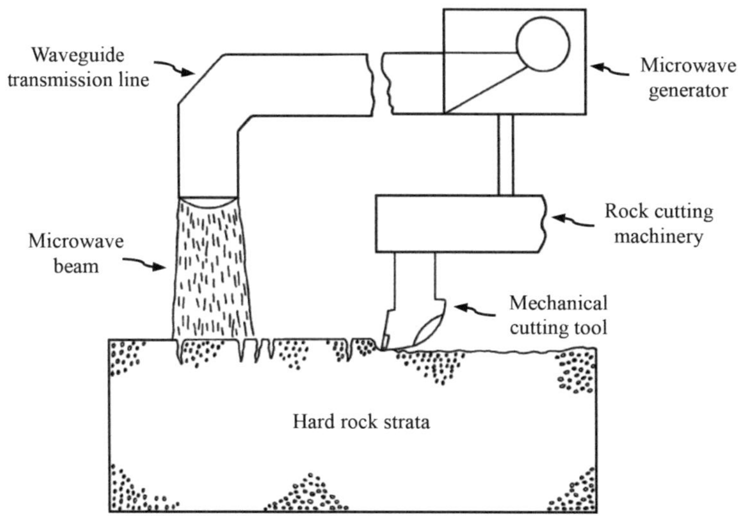

Figure 1.
Schematic diagram of microwave-assisted mechanical rock cutting [4].

rock, breaking due to various advantages including high efficiency and having no dust or noise pollution [1–3]. Realizing microwave-assisted mechanical rock cutting (**Figure 1**) using the microwave-induced hard rock fracturing technique can prolong the mechanical life and improve the efficiency of rock breaking operations [4–7]. At present, tunnel boring machines (TBMs) and shield machines have been increasingly used in tunnel excavation. The shield machine is subjected to a series of problems such as deformation of the tool apron and severe cutter wear due to the presence of boulders [8–10]. During the tunneling of hard rocks by TBMs, the disc cutter is worn and frequently changed-out, thus increasing the cost of maintenance and influencing construction progress [11–14].

The design of cutter heads of shield machines and TBMs is closely related to the properties of rocks and prevailing geological conditions [15, 16]. The mechanical strengths (uniaxial compressive strength, tensile strength, and point load strength, etc.) of rocks are important parameters influencing the service life and penetration of disc cutters on TBMs [17, 18]. Microwave treatment can significantly decrease the strength of rocks [3, 19–22], and thus can improve the penetration and life of disc cutters. Therefore, by introducing microwave heating technology into TBMs or shield machines, hard rocks or boulders can be pre-fractured through microwave irradiation. In this way, cutter wear can be reduced to increase efficiency in tunnel excavation.

Some scholars have carried out numerous experiments and numerical research into the mechanism governing the microwave-induced fracturing of rocks (or ores) and the influence of microwave treatment on the mechanical properties of rocks (or ores) [23–25]. Under the effect of microwave treatment, new intergranular and transgranular fractures are generated in rocks [26] to lead to the reduction of work index [27, 28] and strengths (including uniaxial strength, Brazilian splitting strength, and point load strength) of rocks [2, 29, 30]. More seriously, rocks are cracked and crushed or molten to cause rocks (or ores) to lose all of their bearing capacity [29, 31]. Hassani et al. [3, 6, 32] and Nekoovaght et al. [33, 34] studied the influence of microwave power and irradiation time on the strength of different kinds of rocks by using a frequency of 2450 MHz multi-mode cavity. In addition, they also studied the influence of the distance between the microwave antenna and the rock on the heating characteristics by experiments and numerical comparison.

Peinsitt et al. [35] explored the effects of microwave irradiation on the strength, acoustic velocity, and heating effect of three types of dry and saturated rocks using a 2450 MHz multi-mode cavity. By using an open-ended waveguide set-up at a frequency of 2,450 MHz, Hartlieb et al. [36–38] explored the failure mechanism and thermal physical characteristics of different types of rocks. Lu et al. [29, 39, 40] studied the mechanism, temperature distribution, and crack propagation of microwave-heated rocks using a multi-mode resonator.

At present, the technology of microwave-induced fracturing of hard rock is still in the laboratory research stage. By summarizing the research results of relevant scholars in this field, this work generalizes the mechanism of microwave heating of rock, microwave heating system, heating characteristics, and the effect of microwave heating on rock cracking and mechanical properties.

2. Principle of microwave heating rocks

When a dielectric material is subjected to an alternating current, it absorbs electrical energy, which is dissipated in the form of heat (the dielectric loss). The dielectric constant of the material consists of the real part and the imaginary part, as shown below

$$\varepsilon = \varepsilon' - j\varepsilon'' \tag{1}$$

where the real part (ε') is known as the dielectric constant. The imaginary part (ε'') is known as the loss factor [41].

The loss tangent ($\tan\delta$) is the ratio of the imaginary part (ε'') to the real part (ε'), i.e.

$$\tan\partial = \frac{\varepsilon''}{\varepsilon'} \tag{2}$$

It measures the ability of the dielectric to store energy and convert it into heat.

The microwave absorption capacity of electrolyte material is related to its dielectric properties. The microwave heating mechanism of minerals and rocks in the electromagnetic field is usually expressed by the power density, which can be expressed by the following Equation

$$P = 2\pi\varepsilon_0\varepsilon'' E^2 f \tag{3}$$

where P is the loss power density deposited within the sample; E is the electric field and f is the microwave frequency; ε_0 is the dielectric constant of free space (8.85×10^{-12} F/m) [42].

The temperature of the dielectric material increases when it absorbs microwave energy [41]. According to the laws of thermodynamics the amount of energy required to increase the temperature of a material to a given amount is calculated by the following equations

$$Q = Cm\Delta T \tag{4}$$

$$P = \rho C \frac{\Delta T}{\Delta t} \tag{5}$$

where Q is the energy absorbed by the material; C is the specific heat capacity; m is the mass; ΔT is the temperature increase after absorbing energy; Δt is the time difference.

By combining Eqs. (3) and (5), the rate of heating may be given by

$$\frac{\Delta T}{\Delta t} = \frac{2\pi\varepsilon_0\varepsilon'' E^2 f}{\rho C} \tag{6}$$

An increase in material temperature causes the volume of the material to increase

$$V(T) = V_0(1 + \alpha\Delta T) \tag{7}$$

where, V_0 is the volume at some reference temperature, and α is the coefficient of thermal expansion of the medium [43].

After being irradiated by microwaves, rocks absorb electromagnetic energy that is transformed into thermal energy, causing the temperature rising of rocks. After microwave irradiation, the temperature of the rocks is not uniform, resulting in uneven thermal expansion in the rocks. Different minerals within rocks have different dielectric properties, leading to different rocks have different microwave absorption capacities. Different minerals also have different thermal expansion coefficients, so the thermal expansion property is different after heating. Therefore, due to the different microwave sensitivity and thermal expansibility of different types of rocks, different types of rocks show different heating characteristics and fracturing effects.

3. Microwave heating system for rock

The microwave heating equipment includes a power supply, a magnetron, an isolator, a coupler, an impedance tuner, a rectangular waveguide, a microwave applicator, and a shielded cavity (**Figure 2**). The power supply is used to convert alternating-current into direct-current to thus create conditions for the operation

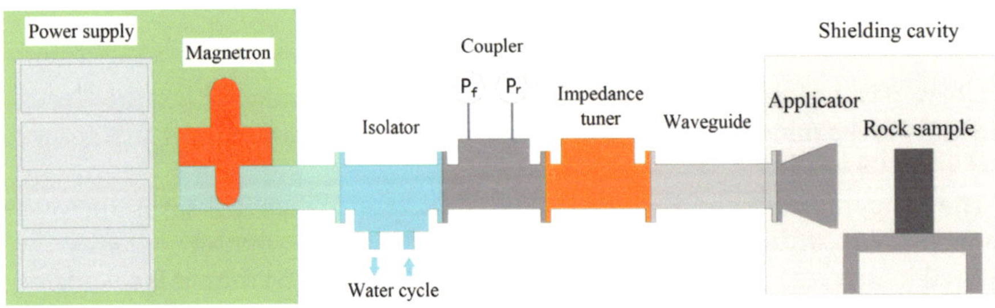

Figure 2.
Schematic view of the microwave heating system.

of the magnetron. The magnetron converts direct-current electrical energy into microwave energy, thus providing continuous microwave power. The isolator is used for the unidirectional circular transmission of microwave energy. The function of the water cycle is to absorb the reflected microwave energy, thus protecting the magnetron from damage. The microwave applicator is used to emit microwave energy to the surface of rocks, where it is used to heat and crack rocks. The impedance tuner is used for impedance matching. Compared with the microwave source with a frequency of 915 MHz, the microwave source with a frequency of 2450 MHz has higher heating efficiency and smaller volume, which is conducive to the combination with the mechanical rock-breaking device. During testing, a metal net is used as shielding to avoid microwave interference with signal transmission to/from the other apparatuses.

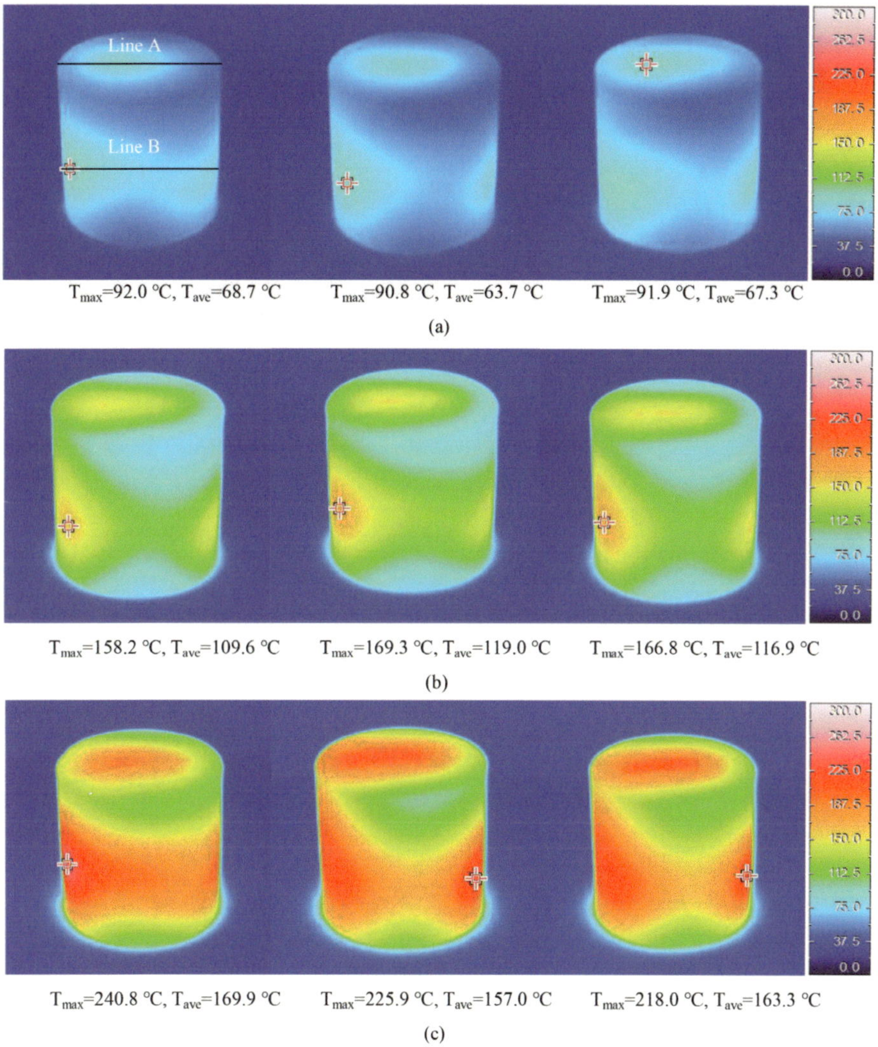

Figure 3.
Temperature distribution on the surface of samples measured by infrared camera (ambient temperature at 13.5°C) [44]. T_{max} = 92.0°C, T_{ave} = 68.7°C T_{max} = 90.8°C, T_{ave} = 63.7°C T_{max} = 91.9°C, T_{ave} = 67.3°C (a) 5 kW, 10 s. T_{max} = 158.2°C, T_{ave} = 109.6°C T_{max} = 169.3°C, T_{ave} = 119.0°C T_{max} = 166.8°C, T_{ave} = 116.9°C (b) 5 kW, 20 s. T_{max} = 240.8°C, T_{ave} = 169.9°C T_{max} = 225.9°C, T_{ave} = 157.0°C T_{max} = 218.0°C, T_{ave} = 163.3°C. (c) 5 kW, 30 s.

4. Temperature characteristics

Figures 3 and **4** show the temperature distributions on the surface of samples and along their straight lines [44]. As is seen, the surface temperature of samples was non-uniformly distributed, i.e. the longer the irradiation time, the more non-uniform the temperature distribution. A local high-temperature region occurred on the surface of the cylindrical samples. The temperature inside the samples was higher than that on the surface of the samples, and the maximum temperature occurred near the center of the samples. There were two high-temperature zones at the left and right sides of the lower part of the cylinder sample surface. The non-uniformity of temperature distribution led to the non-uniformity of thermal expansion within the samples, which will promote the cracking and fracturing of samples. The influence of irradiation time on the surface temperature of the basalt samples is shown in **Figure 5**. As is seen, after microwave irradiation for 10 s, 20 s, and 30 s, the maximum temperatures were 91.6°C, 164.8°C, and 228.2°C at a microwave power of 5 kW while the average temperatures were 66.6°C, 115.2°C, and

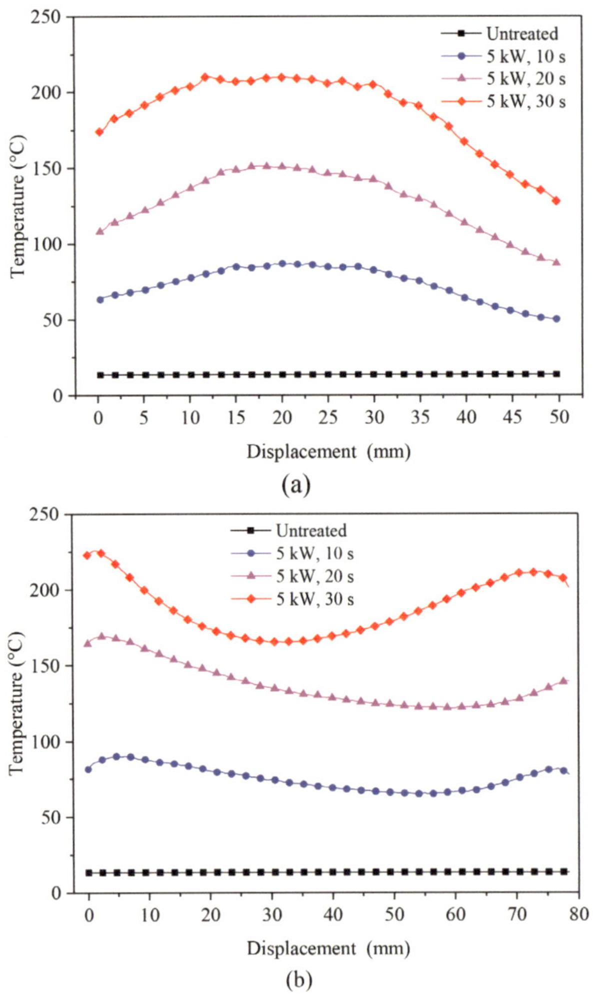

Figure 4.
Temperature distributions along straight lines of samples passing the point with the highest temperature [44]. (a) the upper surface (line a). (b) Cylindrical surface (line B).

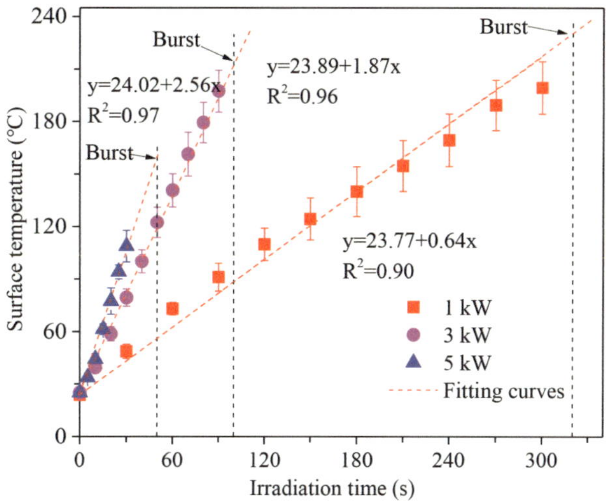

Figure 5.
Surface temperature vs. irradiation time of Φ50 × 100 mm cylindrical basalt specimens [29].

163.4°C, respectively. When the microwave power is constant, the maximum and average temperature on the surface of the samples both linearly increased with the irradiation time (**Figure 5**). The longer the microwave irradiation time, the higher the surface temperature of the samples.

The surface temperatures of the cylindrical basalt samples increased with the increase of irradiation time at each of the three microwave powers (**Figure 5**) [29]. The surface temperature of the sample increased linearly with the microwave irradiation time, and the heating rate increased with the growth of microwave power. Samples burst at approximately 230°C and 320 s at 1 kW, 210°C and 100 s at 3 kW, and 160°C and 50 s at 5 kW. The broken pieces began to melt with increasing the irradiation time. The higher the microwave power input, the faster the rate of heating, and the shorter the time needed for the sample to bursting.

5. Effect of microwave heating on crack propagation of rock samples

5.1 Microscopic crack propagation

Under the conditions of 5 kW power and different microwave irradiation times, the crack propagation on the surface of the 50 mm cubic samples was observed with an ultra-depth field microscope, as shown in **Figure 6**. Compared with the untreated cubic samples, more intragranular and intergranular cracks were observed around and along the olivine granules. With the increase of microwave irradiation time, cracks became wider and more pronounced. Relative to untreated samples, intergranular and intragranular microscopic cracks occurred within cylindrical samples after microwave treatment for 60 s at 3 kW.

After microwave irradiation, more intergranular and transgranular cracks occurred within the basalt samples. In particular, the intergranular cracks mainly occurred between plagioclase and olivine, while the intragranular cracks mainly occurred within the olivine grains. With the increase of intergranular cracks and intragranular cracks, macroscopic cracks mainly occurred across the area where olivine and enstatite grains gathered. This is because enstatite provides the energy needed for the thermal expansion of olivine [29, 39].

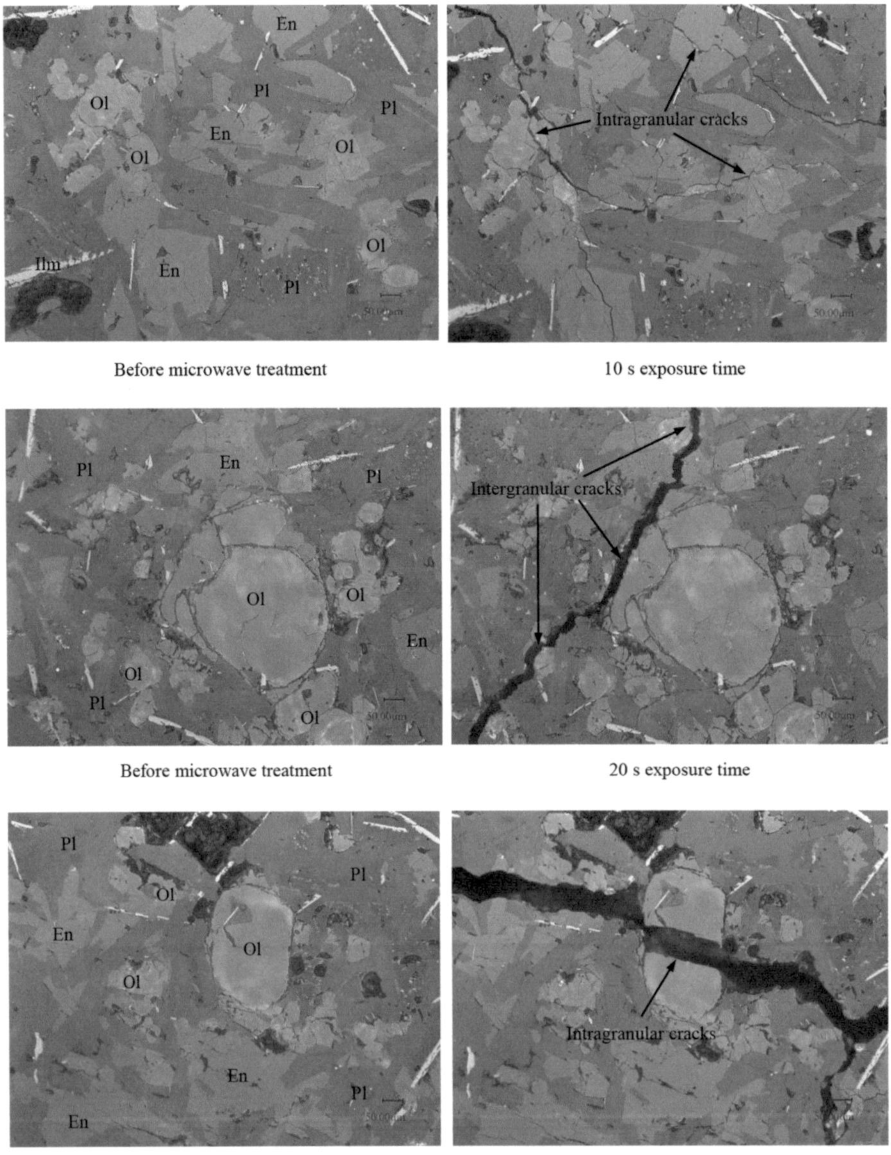

Figure 6.
Microscopic crack propagation on the surface of 50 mm cubic basalt specimens viewed under an ultra-depth of field microscope at 300× magnification before (left) and after (right) microwave treatment at 5 kW power and three exposure times [39].

5.2 Macroscopic crack propagation

The typical pattern of macroscopic crack propagation on the surface of the cylindrical samples is shown in **Figure 7** [29]. It can be seen that the macroscopic cracks of the microwave-treated cylindrical samples propagated on the two end surfaces and the cylindrical curved surface of the cylindrical samples. The main crack on the cylinder was approximately parallel to the bus of the cylinder and connected to the crack on the end surfaces. After a long time of microwave irradiation, about three cracks propagated on the end surfaces of the sample, converging to a point near the center of the circle. The sample with low microwave power or short irradiation time had less crack propagation. With the increase of microwave irradiation time or microwave

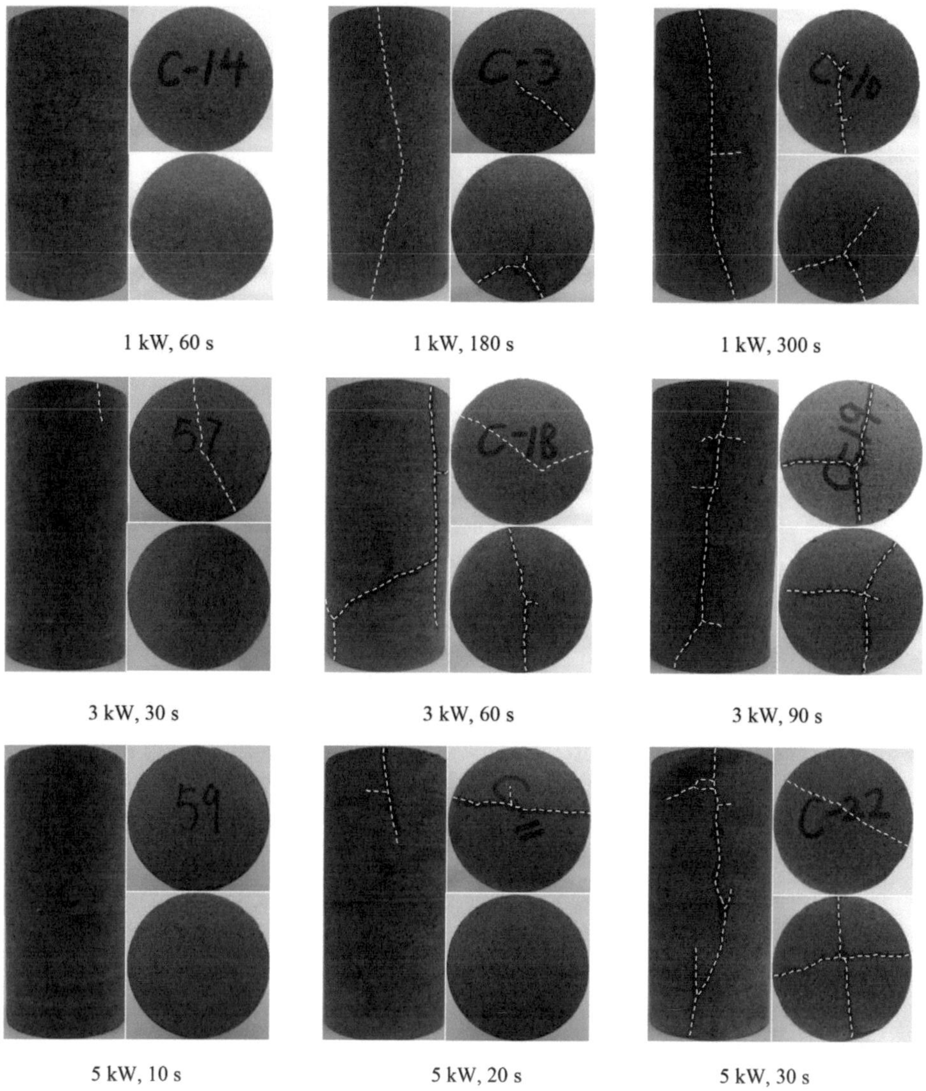

Figure 7.
Typical pattern of macroscopic crack propagation on the surface of cylindrical specimens induced by microwave irradiation at three power levels and seven exposure times [29].

power, the length and number of cracks increased gradually. Microwave power and irradiation time are important parameters affecting crack generation. The higher the microwave power, the shorter the time needed to generate cracks of the same degree.

6. Effect of microwave heating on mechanical properties of rock

6.1 Mechanical strength

After microwave irradiation, the uniaxial compression strength decreases with the increase of microwave irradiation time at three microwave powers (1 kW, 3 kW, and 5 kW). The uniaxial compression strength decreases in an approximately linear manner with increasing microwave irradiation time. The greater the applied microwave power, the faster the decrease rate of uniaxial compression strength [29]. Pyroxene is a highly microwave-absorbing mineral and has strong heating ability after microwave irradiation. Olivine is a strong thermal expansion mineral, which can produce strong thermal expansion under high temperatures. As a result, transgranular cracks mainly occurred in olivine particles, and intergranular cracks mainly occurred between olivines and plagioclases. With the increase of microwave irradiation time, the microcracks slip and connect, forming weak planes in the rocks. Compared with the rocks not irradiated by microwave, the strength of the rocks decreased to different extents. The longer the microwave irradiation time or the higher the microwave power, the more developed the weak surface and the greater the strength reduction.

Figure 8 shows the relationships between the mechanical strength of a basalt and microwave irradiation time for different microwave power levels. As illustrated in **Figure 8(a-c)**, the uniaxial compression strength (a), the Brazilian tensile strength (b), and the point load strength (c) decreased with irradiation time at each power level. Overall, the magnitude of the reductions was least for uniaxial compression strength, and most for point load strength. According to the slope of the fitting curves, the higher the microwave power, the faster the rock strength decreased. For example, the Brazilian tensile strength was reduced by approximately 39% at 1 kW power after 300 s irradiation, 37% at 3 kW power after 90 s irradiation, and 46% at 5 kW power after 30 s irradiation. Similarly, the uniaxial compression strength was reduced by less than 11% at 3 kW power and by 27% at 5 kW power after 30 s irradiation. Under the three microwave powers, Uniaxial compressive, Brazilian tensile, and point load strength were reduced by up to 37%, 46% and 62% respectively. Microwave power and irradiation time are important parameters affecting basalt strength, and the reduction of strength has an approximately linear relationship with the irradiation time at each power level.

At the applied power of 5 kW, the relationship between conventional triaxial compressive strength of basalts and microwave irradiation time is shown in **Figure 9** [45]. Under four confining pressures (σ_3 = 0 MPa, 10 MPa, 30 MPa, and 50 MPa), the conventional triaxial compressive strength reduces at different rates with increasing microwave irradiation time. At 30 s irradiation, the conventional triaxial compressive strength reduces by 27%, 7%, 2%, and 1% under the four confining pressures, respectively. It is worth noting that, with the increase in confining pressure, the conventional triaxial compressive strength of the basalts gradually reduces (The test value has a certain discreteness, and the overall effect is reduced.). Confining pressure closes part of the cracks induced by microwave irradiation and increases the frictional forces that prevent slippage of microcracks. Therefore, confining pressure inhibits the reduction of conventional triaxial compressive strength and discreteness. The discreteness of conventional triaxial compressive strength is mainly caused by the heterogeneity of microcracks, while the confining pressure leads to the closure of microcracks in the samples and reduction of microcracks that produce slippage. Therefore, the confining pressure can decrease the discreteness of conventional triaxial compressive strength. Geological factors such as in situ stress should be considered when microwave-induced fracturing is used in underground geotechnical engineering [45].

Due to rocks are heterogeneous materials, differences between samples can lead to the discreteness of test results. The existence of microwave sensitive minerals and strong thermal expansion minerals results in the random distribution of microcracks in different directions within the rocks after microwave irradiation. After microwave irradiation, microcracks occurred within the rock, which can further increase the dispersion of the test results. The higher the microwave power

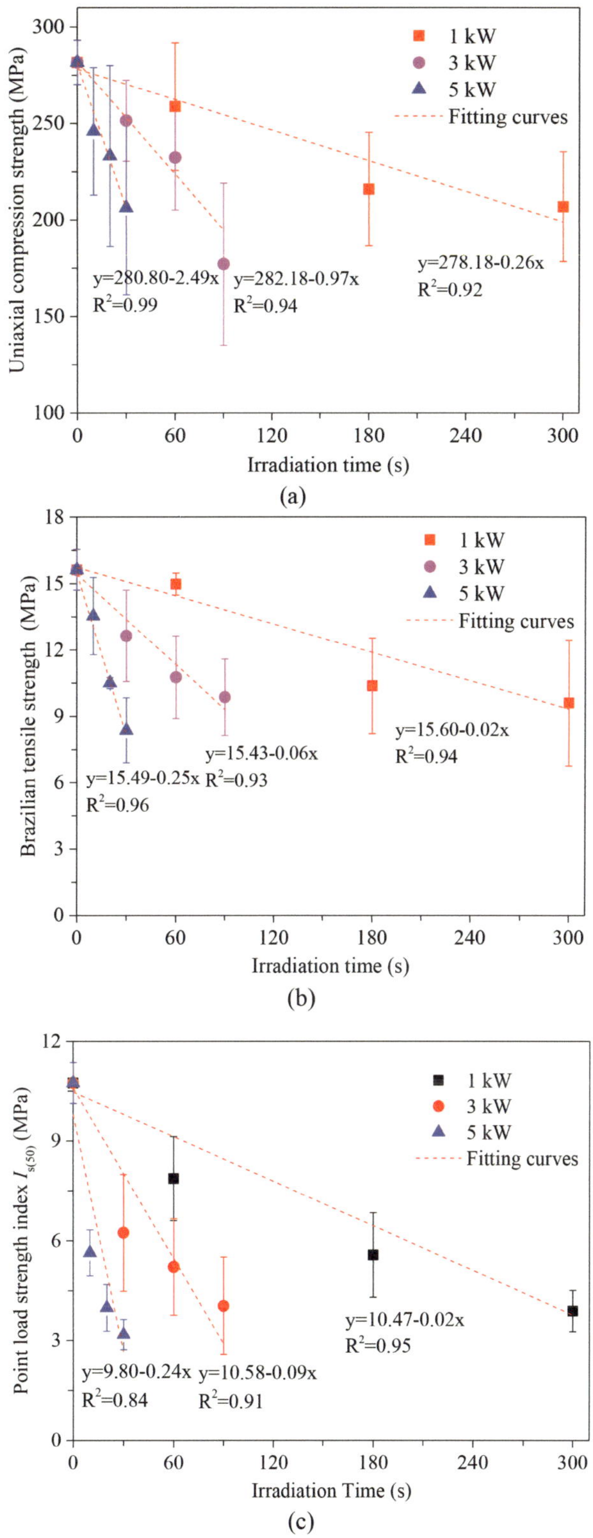

Figure 8.
Relationships between mechanical strength of basalt and microwave irradiation time for three power levels: (a) uniaxial compressive strength, (b) Brazilian tensile strength, and (c) point load strength [29].

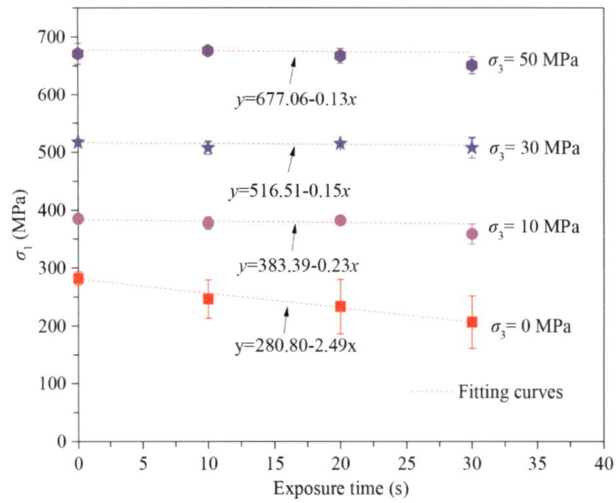

Figure 9.
Reductions in compression strength of basalts under different confining pressures [45].

and the longer the irradiation time, the more discreteness the test results will be. When significant crack propagation or weak surface occurs in the rock, the bearing capacity of the rock will be significantly reduced, leading to the decrease of the rock strength. It is revealed that the confining pressure inhibits discreteness of basalt strength and the strength differences induced by microcracking gradually decrease with increasing confining pressure.

6.2 Elastic modulus and Poisson's ratio

The average elastic modulus and Poisson's ratio are calculated by linear fitting of the stress–strain curve. The elastic modulus and Poisson's ratio are calculated by linear fitting of the stress–strain curve, and the calculation results are shown in **Figures 10** and **11**. The basalt has a compact structure and the average elastic modulus before microwave irradiation is 97 GPa. After microwave irradiation,

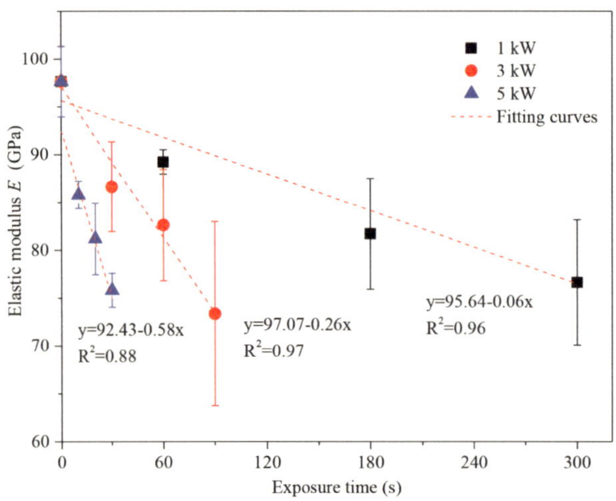

Figure 10.
Relationships between elastic modulus of basalts and microwave exposure time under uniaxial compression [45].

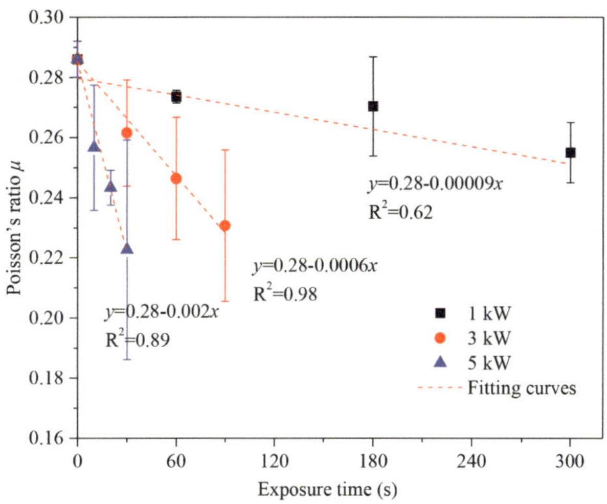

Figure 11.
Relationships between Possion's ratio of basalts and microwave exposure time under uniaxial compression [45].

the elastic modulus and Poisson's ratio both decrease at three power levels (1 kW, 3 kW, and 5 kW). The decrease rate of elastic modulus is greater than Poisson's ratio at the three power levels, and elastic moduli are decreased by 22%, 25%, and 22%, while the Poisson's ratios decrease by 11%, 19%, and 22%, respectively. The elastic modulus and Poisson's ratio decrease linearly with the microwave irradiation time. According to the slopes of the fitting curves, it is known that the higher the microwave power, the greater the degree of reduction of elastic modulus and Poisson's ratio.

After microwave treatment, the decrease of elastic modulus indicates that microwave irradiation can reduce the stiffness of rocks, that is to say, it can reduce the bearing capacity of rocks. As is known, in the compression test, the elastic deformation of rocks is mainly determined by the skeleton of mineral particles. Transgranular fracture and intergranular cracks can be induced in and between mineral particles by microwave treatment. With the increase and penetration of these microcracks, new cracks will be formed. Therefore, microwave irradiation changes the skeleton structure of mineral particles in the rocks and weakens the resistance of the rocks to elastic deformation.

6.3 Cohesion and internal friction angle

In the field of rock mechanics, cohesion refers to the attraction between molecules on the surface of adjacent mineral particles. After microwave treatment, the peak strength of the samples at different irradiation times presented monotonically increasing relation with confining pressure, which was in accordance with the Coulomb strength criterion. The cohesion c and internal friction angle φ of basalt samples were calculated according to the Coulomb strength criterion (**Table 1** and **Figure 12**). After microwave treatment, the cohesion of the samples decreased linearly with the increase of microwave irradiation time. When the microwave power is 5 kW and the microwave irradiation time is 10 s, 20 s, and 30 s, the cohesion decreases by 14%, 13%, and 25% respectively. After microwave treatment, transgranular and intergranular cracks were generated within and between mineral particles, which reduced the cementation between mineral particles and thus reduced the overall cohesion. While the internal friction angle increases slightly after microwave treatment.

Exposure time (s)	σ_3 (MPa)	σ_1 (MPa)	c (MPa)	φ (°)
0	0	281.7	54.26	50.03
	10	385.33		
	30	518		
	50	671		
10	0	245.9	46.41	51.59
	10	377.67		
	30	507.33		
	50	676.33		
20	0	233.1	47.09	51.63
	10	382		
	30	514.67		
	50	667		
30	0	206.3	40.93	52.18
	10	358.67		
	30	507.33		
	50	650.67		

Table 1.
Conventional triaxial compressive strength test results of basalt samples at different exposure times [45].

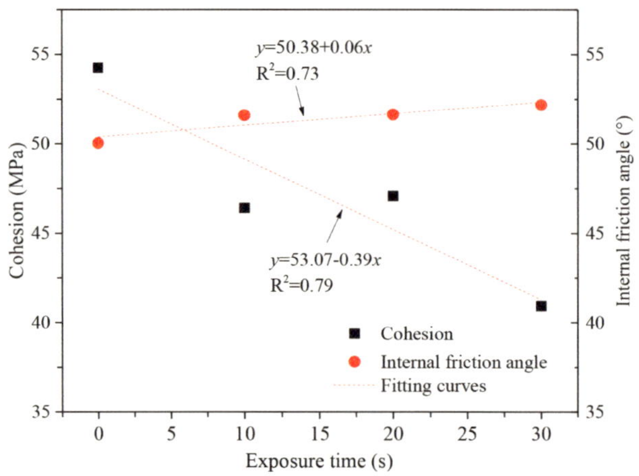

Figure 12.
Relationships between cohesion and internal friction angle of the basalts with microwave exposure time [45].

7. Summary and conclusions

Microwave power level had a significant positive relationship with basalt heating rate: the higher the applied power, the faster the basalt temperature rises. The surface temperature of the sample is not uniformly distributed, which is conducive to the generation of cracks. The higher the microwave power, the more serious the crack propagation. As the number of fractures increased, visible cracks were generated, leading to significant strength reduction. The Uniaxial compressive, Brazilian tensile, and point load strength all decreased after irradiation at each of the three power levels; the higher the power level, the faster the strength decreased. Under the three microwave powers, Uniaxial compressive, Brazilian tensile, and point load strength were reduced by up to 37%, 46%, and 62% respectively. The conventional triaxial compressive strength decreased linearly with microwave irradiation time, and the higher the confining pressure, the smaller the reduction in the strength of basalt samples after microwave treatment. At 30 s irradiation, the conventional triaxial compressive strength reduces by 27%, 7%, 2%, and 1% under the four confining pressures (σ_3 = 0 MPa, 10 MPa, 30 MPa, and 50 MPa), respectively.

The elastic modulus and Poisson's ratio of basalts also decreased in a quasi-linear manner with elapsed microwave irradiation time. At the three applied powers, the decrease rate in elastic modulus always exceeds that of the Poisson's ratio, and elastic moduli are decreased by 22%, 25%, and 22%, while the Poisson's ratios decrease by 11%, 19%, and 22%, respectively. The confining pressure results in the closure of the microcracks caused by microwave irradiation, so the influence of microwave treatment on strength and deformation is reduced, leading to a decrease in the influence on the elastic constants. The cohesion decreases with the increase of microwave irradiation time and presents an approximate linear decreasing relationship. At microwave power of 5 kW and irradiation times of 10 s, 20 s, and 30 s, the cohesion is reduced by 14%, 13%, and 25%, respectively. In the basalt samples, new microcracks in various directions generated by microwave irradiation can increase the discreteness of test results, while the discreteness of test results caused by microcracks gradually reduces with increasing confining pressure.

Uniaxial compressive, Brazilian tensile, and point load strength are important parameters that affect the service life and the penetration of mechanical rock-breaking tools. The reduction of rock strength can increase the service life and the penetration of mechanical rock-breaking tools. Microwave irradiation weakens the mechanical properties of rock, which effectively reduces the resistance of the rock to the mechanical rock-breaking tool, which can reduce the wear of the mechanical rock-breaking tool and improve the rock-breaking efficiency. Microwave-assisted rock-breaking has significant potential application to in-situ mining, tunneling, rock breakage, and comminution.

Acknowledgements

Financial support for this work by the National Natural Science Foundation of China (No: 42002281), by the Natural Science Foundation of Henan Province (No: 202300410002, 212300410325), by the Science and Technology Research and Development Plan of China Railway Group Limited (No: 2020-Zhongda-06) and the Science and Technology Innovation Plan of China Railway Tunnel Group (No: Suiyanhe 2020-11) are greatly appreciated.

Author details

Gaoming Lu[1,2*] and Jianjun Zhou[1,2]

1 State Key Laboratory of Shield Machine and Boring Technology, Zhengzhou, China

2 China Railway Tunnel Group Co., Ltd, Guangzhou, China

*Address all correspondence to: gaoming_lu@foxmail.com

References

[1] Lu GM, Feng XT, Li YH, Zhang XW. The Microwave-Induced Fracturing of Hard Rock[J]. Rock Mechanics and Rock Engineering, 2019, 52(9): 3017-3032.

[2] Lu GM. Experimental study on the microwave fracturing of hard rock[D]. Northeastern University, 2018.

[3] Hassani F, Nekoovaght PM, Gharib N. The influence of microwave irradiation on rocks for microwave-assisted underground excavation[J]. Journal of Rock Mechanics and Geotechnical Engineering, 2016, 8(1): 1-15.

[4] Lindroth DP, Morrell RJ, Blair JR. Microwave assisted hard rock cutting: US5003144[P]. 1991.

[5] Feng XT, Lu GM, Li YH, Zhang XW. Cutter head for microwave presplitting type hard-rock tunnel boring machine: US10428654B2[P]. 2019.

[6] Hassani F, Nekoovaght P. The development of microwave assisted machineries to break hard rocks[C]// Proceedings of the 28th International Symposium on Automation and Robotics in Construction (ISARC), Seoul, 2011: 678-684.

[7] Ouellet J, Radziszweski P, Raghavan V, Hemanth S. Electromagnetic energy assisted drilling system and method: US8550182B2[P]. 2019-1-1.

[8] Nishitake S. Earth pressure balanced shield machine to cope with boulders[J]. International Journal of Rock Mechanics and Mining Sciences & Geomechanics Abstracts, 1989, 1(6): 552-572.

[9] Filbà M, Salvany JM, Jubany J, Carrasco L. Tunnel boring machine collision with an ancient boulder beach during the excavation of the Barcelona city subway L10 line: A case of adverse geology and resulting engineering solutions[J]. Engineering Geology, 2016, 200(2): 31-46.

[10] Li XG, Yuan DJ. Creating a working space for modifying and maintaining the cutterhead of a large-diameter slurry shield: A case study of Beijing railway tunnel construction[J]. Tunnelling and Underground Space Technology, 2018, 72(2): 73-83.

[11] Jain P, Naithani AK, Singh TN. Performance characteristics of tunnel boring machine in basalt and pyroclastic rocks of Deccan traps–A case study[J]. Journal of Rock Mechanics and Geotechnical Engineering, 2014(1): 36-47.

[12] Entacher M, Lorenz S, Galler R. Tunnel boring machine performance prediction with scaled rock cutting tests[J]. International Journal of Rock Mechanics and Mining Sciences, 2014, 70(9): 450-459.

[13] Xia YM, Zhang K, Liu JS. Design optimization of TBM disc cutters for different geological conditions[J]. World Journal of Engineering and Technology, 2015(4): 218-231.

[14] Rostami J. Performance prediction of hard rock Tunnel Boring Machines (TBMs) in difficult ground[J]. Tunnelling and Underground Space Technology, 2016, 57: 173-182.

[15] Xia YM, Ouyang T, Zhang XM, Luo DZ. Mechanical model of breaking rock and force characteristic of disc cutter[J]. Journal of Central South University, 2012, 19(7): 1846-1858.

[16] Deliormanlı AH. Cerchar abrasivity index (CAI) and its relation to strength and abrasion test methods for marble stones[J]. Construction and Building Materials, 2012, 30: 16-21.

[17] Wijk G. A model of tunnel boring machine performance[J]. Geotechnical & Geological Engineering, 1992, 10(1): 19-40.

[18] Boniface A. Tunnel boring machine performance in basalts of the Lesotho formation[J]. Tunnelling and Underground Space Technology, 2000, 15(1): 49-54.

[19] Lu GM, Li YH, Hassani F, Zhang XW. Review of theoretical and experimental studies on mechanical rock fragmentation using microwave-assisted approach[J]. Chinese Journal of Geotechnical Engineering, 2016, 38(8): 1497-1506.

[20] Zheng YL, Ma ZJ, Zhao XB, He L. Experimental investigation on the thermal, mechanical and cracking behaviours of three Igneous rocks under microwave treatment[J]. Rock Mechanics and Rock Engineering, 2020.

[21] Zheng YL, Zhang QB, Zhao J. Effect of microwave treatment on thermal and ultrasonic properties of gabbro[J]. Applied Thermal Engineering, 2017, 127: 359-369.

[22] Li X, Wang S, Xu Y, Yao W, Xia K, Lu G. Effect of microwave irradiation on dynamic mode-I fracture parameters of Barre granite[J]. Engineering Fracture Mechanics, 2019, 224: 106748.

[23] Toifl M, Hartlieb P, Meisels R, Antretter T, Kuchar F. Numerical study of the influence of irradiation parameters on the microwave-induced stresses in granite[J]. Minerals Engineering, 2017, 103-104(4): 78-92.

[24] Ali AY, Bradshaw SM. Confined particle bed breakage of microwave treated and untreated ores[J]. Minerals Engineering, 2011, 24(14): 1625-1630.

[25] Jones DA, Kingman SW, Whittles DN, Lowndes IS. The influence of microwave energy delivery method on strength reduction in ore samples[J]. Chemical Engineering and Processing: Process Intensification, 2007, 46(4): 291-299.

[26] Kingman SW, Corfielf GM, Rowson NA. Effects of microwave radiation upon the mineralogy and magnetic processing of a massive Norwegian ilmenite ore[J]. Magnetic and Electrical Separation, 1998, 9(3): 131-148.

[27] Kingman SW, Vorster W, Rowson NA. The influence of mineralogy on microwave assisted grinding[J]. Minerals Engineering, 2000, 13(3): 313-327.

[28] Vorster W, Rowson NA, Kingman SW. The effect of microwave radiation upon the processing of Neves Corvo copper ore[J]. International Journal of Mineral Processing, 2001, 63(1): 29-44.

[29] Lu GM, Feng XT, Li YH, Hassani F, Zhang X. Experimental investigation on the effects of microwave treatment on basalt heating, mechanical strength, and fragmentation[J]. Rock Mechanics and Rock Engineering, 2019, 52(8): 2535-2549.

[30] Kingman SW, Jackson K, Bradshaw SM, Rowson NA, Greenwood R. An investigation into the influence of microwave treatment on mineral ore comminution[J]. Powder Technology, 2004, 146(3): 176-184.

[31] Li Yuan-hui, Lu Gao-ming, Feng Xia-ting, Zhang Xiwei. The influence of heating path on the effect of hard rock fragmentation using microwave assisted method[J]. Chinese Journal of Rock Mechanics and Engineering, 2017, 36(6): 1460-1468.

[32] Hassani F, Nekoovaght PM, Radziszewski P, Waters KE. Microwave assisted mechanical rock breaking[C]// Proceedings of the 12th ISRM

International Congress on Rock Mechanics, Beijing: International Society for Rock Mechanics, 2011: 2075-2080.

[33] Nekoovaght P, Hassani F. The influence of microwave radiation on hard rocks as in microwave assisted rock breakage applications[M]. Boca Raton: Rock Engineering and Rock Mechanics: Structures in and on Rock Masses, 2014: 195-198.

[34] Nekoovaght P, Gharib N, Hassani F. Numerical simulation and experimental investigation of the influence of 2.45 GHz microwave radiation on hard rock surface[C]// ISRM International Symposium—8th Asian Rock Mechanics, Sapporo, Japan: International Society for Rock Mechanics, 2014.

[35] Peinsitt T, Kuchar F, Hartlieb P, Moser P, Kargl H, Restner U, Sifferlinger N. Microwave heating of dry and water saturated basalt, granite and sandstone[J]. International Journal of Mining and Mineral Engineering, 2010, 2(1): 18-29.

[36] Hartlieb P, Kuchar F, Moser P, Kargl H, Restner U. Reaction of different rock types to low-power (3.2 kW) microwave irradiation in a multimode cavity[J]. Minerals Engineering, 2018, 118: 37-51.

[37] Hartlieb P, Toifl M, Kuchar F, Meisels R, Antretter T. Thermo-physical properties of selected hard rocks and their relation to microwave-assisted comminution[J]. Minerals Engineering, 2016, 91: 34-41.

[38] Hartlieb P, Leindl M, Kuchar F, Antretter T, Moser P. Damage of basalt induced by microwave irradiation[J]. Minerals Engineering, 2012, 31(3): 82-89.

[39] Lu GM, Zhou JJ, Li YH, Zhang X, Gao WY. The influence of minerals on the mechanism of microwave-induced fracturing of rocks[J]. Journal of Applied Geophysics, 2020, 180: 1-11.

[40] Lu GM, Li YH, Hassani F, Zhang X. The influence of microwave irradiation on thermal properties of main rock-forming minerals[J]. Applied Thermal Engineering, 2017, 112(2): 1523-1532.

[41] Chantrey P. Industrial Microwave Heating[M]. London: Peter Peregrinus Ltd, 1983: 1-659.

[42] Meredith R. Engineers' handbook of industrial microwave heating[J]. Engineers' handbook of industrial microwave heating [Book Review], 1998, 13(1): 3-3.

[43] Ahrens TJ. Mineral physics and crystallography: a handbook of physical constants[M]. Washington, dc: American Geophysical Union, 1995.

[44] Lu GM, Sun ZC, Zhou JJ, Chen K, Li FY. Effect of Microwave irradiation on computed tomography and acoustic emission characteristics of hard rock[J]. Geotechnical and Geological Engineering, 2020.

[45] Lu GM, Feng XT, Li YH, Zhang X. Influence of microwave treatment on mechanical behaviour of compact basalts under under different confining pressures[J]. Journal of Rock Mechanics and Geotechnical Engineering, 2020, 12(2): 1-10.

Section 2

Non-Thermal Effect of Electromagnetic Field

Chapter 8

Microwave Heating of Liquid Crystals and Ethanol-Hexane Mixed Solution and Its Features (Review)

Akira Naito, Yugo Tasei, Batsaikhan Mijiddorj, Izuru Kawamura and Kazuyoshi Ueda

Abstract

Microwave heating is widely used to accelerate organic reactions in the chemistry field. However, the effect of microwaves on chemical reaction has not yet been well characterized at the molecular level. In this review chapter, microwave heating processes of liquid crystals and an ethanol-hexane mixed solution under microwave irradiation were experimentally and theoretically investigated using *in situ* microwave irradiation nuclear magnetic resonance (NMR) spectroscopy and molecular dynamics (MD) simulation, respectively. The temperature of the solution under microwave irradiation was estimated from a chemical shift calibrated temperature (CSC-temperature) which was determined from the temperature dependence of the ^{1}H chemical shift. The CSC-temperatures of CH_2 and CH_3 non-polar protons of ethanol reflect the bulk temperature of a solution by the thermal microwave effect. The lower CSC-temperature of the OH polar protons in ethanol and much higher CSC-temperature of H-C=N (7′) and CH_3-O (α’) protons of N-(4-methoxybenzyliden)-4-butylaniline with respect to the bulk temperature are attributed to the non-thermal microwave effects. According to the MD simulation under microwave irradiation, the number of hydrogen bonds increased in the ethanol-hexane mixed solution as a result of a non-thermal microwave effect. It is concluded that a coherently ordered low entropy state of polar molecules is induced by a non-thermal microwave effect. The ordered state induces molecular interaction, which may accelerate the chemical reaction rate between molecules with polar groups.

Keywords: microwave heating mechanism, microwave irradiation NMR spectroscopy, MD simulation, thermal microwave effect, non-thermal microwave effect

1. Introduction

Microwave heating effects in the field of chemical science are attributed to an increase in the solvent temperature due to dielectric loss [1–6]. Dipole moments of the solvent molecules align along an applied electric field that oscillates in the case of microwaves. As the dipoles attempt to align along this alternating electric field, which is a low entropy state, heat energy is produced by molecular friction and collision, which increases entropy and the energy is dissipated through the system. From a thermodynamic point of view, this microwave energy converts to heat energy (thermal effect) and work energy such as volume change (non-thermal effect). However, the detailed molecular mechanisms associated with thermal and non-thermal microwave effects on the chemical reaction rates have not yet been fully elucidated. In particular, the non-thermal microwave effect has not yet been well characterized [7]. In the field of chemistry, microwave heating is widely used to accelerate organic synthesis reactions [2–4, 7–14], reduce polymerization reaction times [15–18], and enhance the activity of enzymes in the field of biological chemistry [19–21]. The majority of accelerated reactions achieved in this manner can be mainly explained by the thermal microwave effect [7, 22]. Non-thermal microwave effects have also been identified and the thermal and non-thermal microwave effects can be distinguished [23]. The non-thermal microwave effects have recently been demonstrated by the observation of an increasing polymerization reaction rate under microwave electric fields and a decrease in the rates under microwave magnetic fields [18]. Nevertheless, non-thermal microwave effects at the molecular level are still controversial [7]. In particular, non-thermal microwave effects have been considered as a direct interaction of the electric field with polar molecules in the reaction medium which is not explained with a macroscopic temperature effect [2, 22]. The presence of an electric field leads to effects on the orientation of dipolar molecules or intermediates, and thus changes the pre-exponential factor A or the activation energy in the Arrhenius equation for certain types of reactions [2, 22].

As a specific microwave heating effect, non-equilibrium localized heating is defined as the generation of isolated regions with a much higher temperature than the bulk solution. This has been observed in liquid–solid systems under microwave irradiation, such as in the case of dimethyl sulfoxide (DMSO) molecules in contact with Co particles under microwave irradiation [24].

To characterize the microwave heating mechanism with the molecular resolution, microwave-irradiation nuclear magnetic resonance (NMR) spectroscopy was first developed by Naito et al. [25]. The characteristic that microwave heating causes a rapid temperature jump was used to obtain state-correlated two-dimensional (2D) NMR spectra between liquid crystalline and isotropic phases. This enabled high-resolution observation of a ^1H-^1H dipolar pattern in the cross section of the ^1H state-correlated 2D NMR spectra of the liquid state toward the liquid crystalline state. The local dipolar interaction of individual protons in the liquid crystalline state can be examined via high-resolution resonance in the isotropic phase [26–30]. The resulting data can also be used to obtain state-correlated 2D NMR spectra of proteins in the native and denatured states [31].

In situ microwave irradiation NMR spectroscopy was developed later [32], and the microwave heating process was observed in liquid crystalline samples [32, 33] and ethanol-hexane mixed solution [34]. The *in situ* temperature of the bulk solution was determined using the relation between ^1H chemical shift and temperature. This temperature is defined as chemical shift calibrated temperature (CSC-temperature) and is measured for individual protons. In N-(4-methoxybenzylidene)-4-butylaniline (MBBA) molecules, H-C=N (7′), and CH_3-O (α’) protons showed significantly higher CSC-temperatures than the bulk temperature in the isotropic state [33]. In the ethanol-hexane mixed solution, OH proton showed lower CSC-temperature than that of the bulk solution [34].

Molecular dynamics (MD) simulations have recently been conducted to investigate the atomic-scale properties of molecular systems under an electric field. Tanaka and Sato investigated the heating process of water and ice under microwave irradiation. They observed that the rotational motion of the water was delayed due to the microwave electric field, and the energy is transferred to the kinetic and intermolecular energies of water [5]. Caleman and van der Spoel described that an infrared (IR) laser pulse energy rapidly increased the intramolecular bond vibrations, and the energy was transferred to rotational and translational motion of ice using MD simulations [35]. Marklund et al. investigated the different orientations of a protein depending on the range of electric field strengths without loss of the structure [36]. English and MacElroy conducted MD simulations and found that microwave heating was more efficient for polarizable water models than for non-polarizable models. [37, 38].

This chapter describes the microwave heating processes for common organic solvents, a mixed solution of ethanol (polar molecule) and hexane (non-polar molecule), and liquid crystalline and isotropic phases of MBBA systems. CSC-temperature was employed to accurately measure the temperature of the bulk solution during microwave irradiation [39]. In the case of diamagnetic nuclei, it has been reported that the temperature dependence of the chemical shift values is typically linear [40–44]: therefore, the CSC-temperatures of the individual groups of molecules under microwave irradiation were assessed using *in situ* microwave irradiation NMR spectroscopy.

2. *In situ* microwave irradiation NMR spectrometer

The *in situ* microwave NMR spectrometer employed consisted of a solid-state NMR spectrometer and a microwave generator that was capable of transmitting 1.3 kW continuous wave (CW) and pulsed microwaves at a frequency of 2.45 GHz (**Figure 1A**). This spectrometer enables NMR signals to be obtained without the interference of microwaves, while radio waves and microwave irradiation are

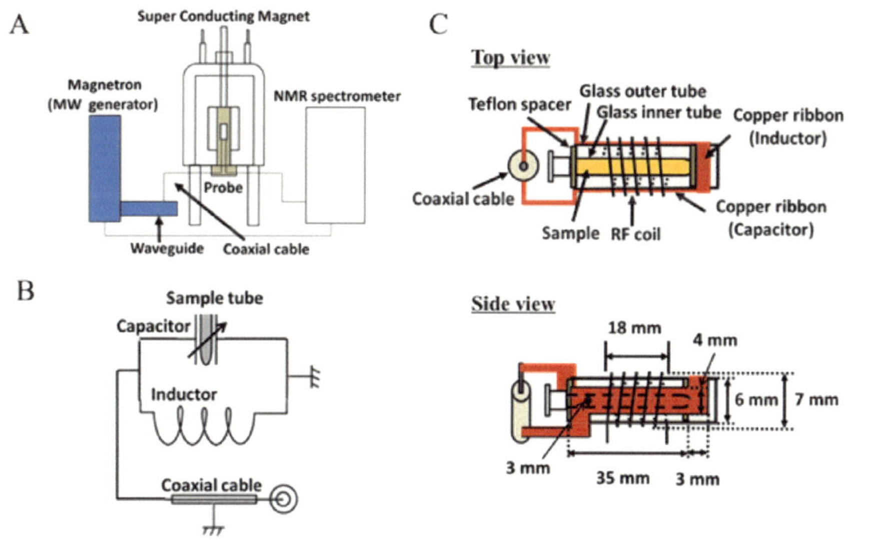

Figure 1.
(A) In situ microwave irradiation solid-state NMR spectrometer which consists of a superconducting magnet, an NMR spectrometer (CMX infinity 400, Chemagnetics), and a microwave transmitter (IDX, 1.3 kW, Tokyo electron Co., ltd.). CW and pulsed microwave are gated by the pulse programmer of the NMR spectrometer. Microwaves are generated and transmitted through a waveguide and coaxial cable, and finally introduced to the microwave resonance circuit in the probe. (B) Equalizing microwave resonance circuit consisting of an inductor and capacitor. The sample tube was located inside the capacitor. (C) Schematic diagrams of microwave resonance and radio wave resonance circuits. The top view indicates the names of components and the side view shows the dimensions of the components. Adapted with permission from [32]. Copyright (2015) Elsevier Inc.

simultaneously applied. Therefore, *in situ* NMR observation is possible under microwave irradiation using this *in situ* microwave irradiation NMR spectrometer. A 3 mm wide flat copper ribbon was used to form the capacitor of the microwave resonance circuit (**Figure 1B**) and was wound coaxially inside the radio wave circuit to increase isolation during microwave irradiation, which reduced interference between microwaves and radio waves (**Figure 1C**) [32]. The microwave resonance circuit was tuned to 2.45 GHz and the radio wave circuit was set to 398 MHz for ^{1}H NMR using a sweep generator. Microwaves were generated from the microwave generator and transmitted through a waveguide and a coaxial cable, and finally, the microwaves were guided to the resonator circuit at the probe head. ^{1}H NMR signals were obtained using 5.0 μs of 90° pulse under microwave irradiation with the *in situ* microwave irradiation NMR spectrometer. The temperature of the sample was varied using the temperature control unit of the spectrometer. Samples were packed in an inner glass tube to insulate them from thermal contact with the outer glass tube. During microwave irradiation, the temperature was controlled at 0°C for the ethanol-hexane mixed solution and at 50°C for MBBA in the isotropic state.

3. MD simulations of ethanol-hexane (1:1) mixed solution and disordered MBBA molecular system

MD simulations of a mixed solution of hexane and ethanol were performed under conditions with and without an oscillating external electric field using Gromacs-2018.7 [45], with the CHARMM36 force field [46]. The CHARMM36 parameter of the MBBA molecule was generated by CHARMM General Force Field (CGenFF) software [47]. Two different systems were applied. The first system consists of 4500 hexane and 4500 ethanol molecules randomly inserted in a cubic box with an edge size of 11.4 nm. The second system includes 904 arbitrarily oriented MBBA molecules in a cubic box with an edge size of 7.5 nm. Periodic boundary conditions were used in all directions, and an oscillating electric field was applied along the x-axis for both systems. An applied electric field with an intensity of 0.5 V/nm and a frequency of 2.45 GHz that served as the microwave heating process [48] was implemented in the simulations. The systems were minimized using steepest descent minimization to reduce steric clashes and were then equilibrated under a constant number of atoms, volume, and temperature (NVT), and under a constant number of atoms, pressure, and temperature (NPT) for each 100-ps MD run. A simulation of the mixture of ethanol and hexane was conducted without the applied electric field for 50 ns. The last snapshot of the simulation was applied as an initial configuration of eight simulations for 5 ns each at different temperatures of 303, 313, 323, and 333 K in the presence and absence of the electric field. In the case of the MBBA system, the initial simulation was performed for 10 ns. The temperature of 293 K (<liquid crystalline to isotropic phase transition temperature (Tc)) and 315 K (>Tc) were then considered with and without the external electric field. The temperature was controlled by a velocity-rescale thermostat [49], and a Parrinello-Rahman barostat provided 1 atm pressure during the simulations [50]. The particle mesh Ewald method [51, 52] and a cutoff of 14 Å were applied for the long-range electrostatic and short-range nonbonded interactions, respectively. The LINCS algorithm was used to constrain all bonds to equilibrium lengths [53]. The time steps of the simulations were 1 and 2 fs for the hexane-ethanol and MBBA systems, respectively. The data were saved at 1 ps intervals. Gromacs tools were used for data analysis; Grace [54] and VMD [55] software were applied for the plots and the structural representations, respectively.

4. *In situ* temperature measurements under microwave irradiation

The *in situ* temperatures of the ethanol-hexane mixed solution and MBBA in the isotropic state under microwave irradiation were assessed with respect to the variation in ^1H chemical shifts with temperature for the sample solution located in the NMR probe. **Figure 2** shows the temperature dependence of the ^1H NMR signals of OH, CH_2, and CH_3 protons in ethanol (**Figure 2a**) and the CH_2 and CH_3 protons in hexane (**Figure 2b**) that were observed under temperature control with the NMR spectrometer. The ^1H chemical shift changes from that at 0°C (Δδ) were plotted as a function of the temperature.

Δδ plotted as a function of temperature was approximately linear within a small temperature range, as observed in methanol and glycol [40]. ^1H NMR spectra of individual protons were measured in the same sample at the same position in the probe and the temperature variation was measured under microwave irradiation. The *in situ* temperatures (CSC-temperature) of the individual protons in the sample solution under microwave irradiation were thus evaluated using the temperature variation of the ^1H chemical shifts. It is stressed that the *in situ* temperature of the bulk solution was accurately determined with the CSC-temperature for individual non-polar proton groups under microwave irradiation.

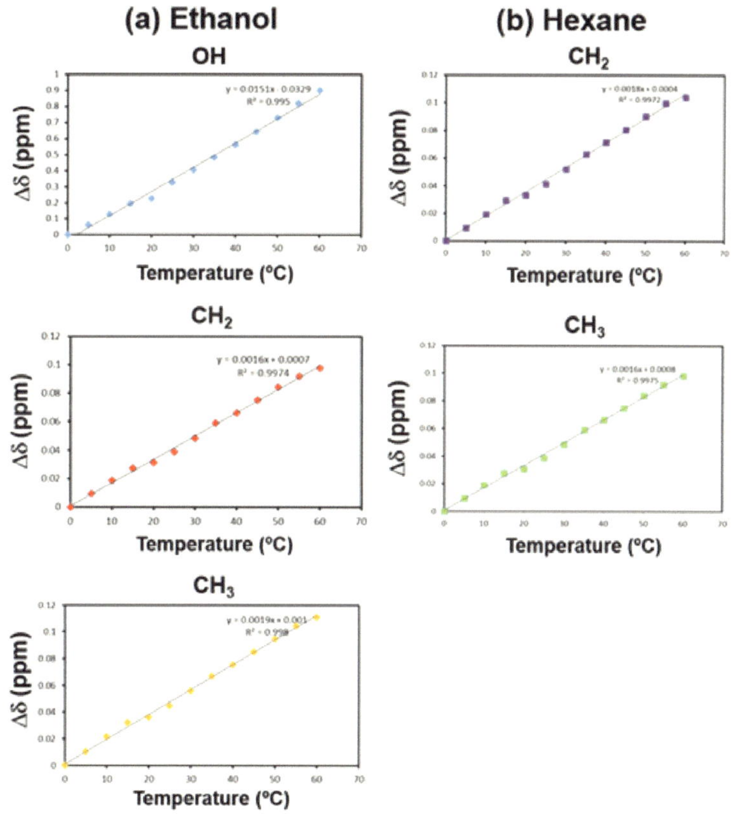

Figure 2.
^1H chemical shift changes from ^1H chemical shift at 0°C (Δδ) for (a) ethanol OH, CH_2, and CH_3 protons and (b) hexane CH_2 and CH_3 protons as a function of temperature in the range from 0-60°C. Adapted with permission from [34]. Copyright (2020) American Chemical Society.

5. CSC-temperature of ethanol-hexane mixed solution

The Δδ values for individual protons of ethanol and hexane changed linearly to the higher field in the range from 0 to 60°C as shown in **Figure 2a** and **b**. Δδ for the OH proton changed 0.9 ppm during the temperature increase from 0 to 60°C, while those for CH_2 and CH_3 changed 0.1 ppm during the temperature increase from 0 to 60°C. Thus, the CSC-temperature under microwave irradiation was accurately determined using these relations for individual protons of a sample solution.

6. Microwave heating process of ethanol, hexane, and ethanol-hexane mixed solution

The microwave heating processes of ethanol were measured by plotting the CSC-temperature increase as a function of the microwave irradiation time, as shown in **Figure 3A(a)**. The temperature was initially set at 0°C, and the samples were continuously irradiated with 135 W (output power of microwave generator) microwave, during which NMR spectra were acquired every 30 s. The CSC-temperature of the CH_2 and CH_3 protons of ethanol increased from 0 to 30°C within 1 min and gradually increased to 58°C under microwave irradiation for 10 min. In contrast, the CSC-temperature of the OH protons increased from 0 to 35°C within 1 min and only slightly increased to 43°C for 10 min. The CSC-temperature of the OH protons deviated to a lower temperature than those of the CH_2 and CH_3 protons by 15°C under microwave irradiation for 10 min.

Figure 3A(b) shows the ^{1}H NMR spectrum of ethanol measured at 55°C (black) and that measured under microwave irradiation (135 W) for 10 min (orange) where the temperature was set at 0°C using the temperature controller of the NMR

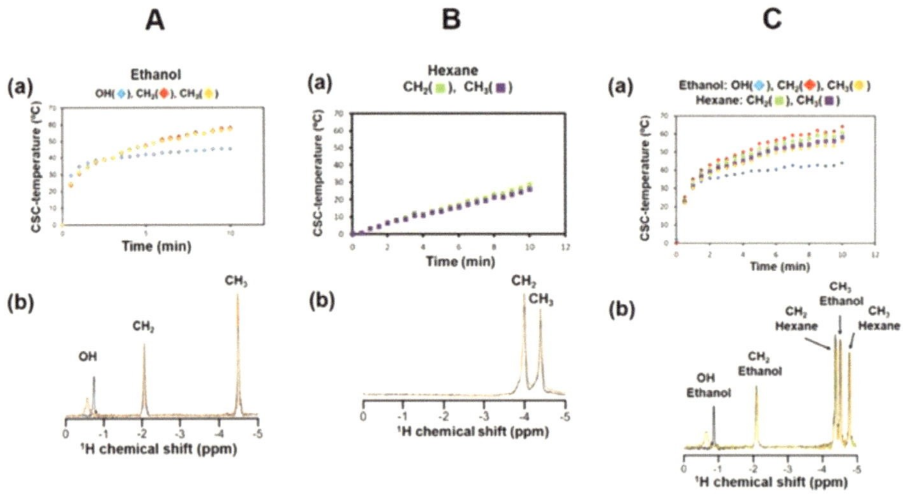

Figure 3.
A(a). CSC-temperatures as a function of microwave irradiation time. CSC-temperatures were determined using the slopes obtained for the individual protons. A(b). ^{1}H NMR spectrum for ethanol regulated at 55°C (black) and that under CW microwave irradiation for 10 min while controlling the instrument temperature setting at 0°C (orange). B(a). CSC-temperatures of CH_2 and CH_3 protons of hexane as a function of microwave irradiation time. B(b). ^{1}H NMR spectra of hexane regulated at 25°C (black) and that under CW microwave irradiation with the same condition as A(b) (orange). C(a). CSC-temperatures of CH_2, CH_3, and OH protons of ethanol and the CH_2 and CH_3 protons of hexane in ethanol-hexane (1:1, v/v) mixed solution as a function of microwave irradiation time. C(b). ^{1}H NMR spectra of ethanol-hexane (1:1, v/v) mixed solution regulated at 55°C (black) and that under CW microwave irradiation with the same condition as A(b) (orange). Adapted with permission from [34]. Copyright (2020) American Chemical Society.

spectrometer. The signal for OH protons under microwave irradiation for 10 min appeared 0.2 ppm lower field (orange). It is noted that the chemical shifts of the CH_2 and CH_3 protons completely overlapped with those measured at 55°C. Therefore, the ^{1}H chemical shifts of the CH_2 and CH_3 protons reflect the bulk temperature of the solution because CH_2 and CH_3 are non-polar groups. The lower field ^{1}H chemical shifts of OH protons under microwave irradiation are thus evidence of being induced by a non-thermal microwave effect in addition to the thermal microwave effect.

Figure 3B(a) shows the CSC-temperatures as a function of the microwave irradiation time. CSC-temperatures of the CH_2 and CH_3 protons of hexane increased gradually to 30°C under microwave irradiation for 10 min. A small temperature increase is attributed to the much lower dielectric loss factor of hexane than that of ethanol.

Figure 3B(b) shows ^{1}H NMR spectra of hexane at 30°C (black) and that after 10 min microwave heating (orange), where the black and orange peaks were completely overlapped.

Figure 3C(a) shows the CSC-temperature of the CH_2, CH_3, and OH protons of ethanol, and the CH_2 and CH_3 protons of hexane in ethanol-hexane (1:1, v/v) mixed solution. The CSC-temperature of the CH_2 and CH_3 protons increased to 40°C within 1 min and gradually increased to 55°C for 10 min. It is noted that all CH_2 and CH_3 protons increased in the same manner. On the other hand, the CSC-temperature of the OH proton was 15°C lower than those of the CH_2 and CH_3 protons.

Figure 3C(b) shows ^{1}H NMR spectra of the ethanol-hexane mixed solution at 55°C (black) and those after microwave heating for 10 min (orange), where the black and orange peaks were almost overlapped. On the other hand, the NMR peak of the OH proton under microwave irradiation appeared 0.2 ppm lower field, which indicates a 15°C lower temperature, as in the case of ethanol.

7. Microwave heating process of MBBA in the liquid crystalline state

Figure 4A shows the molecular structure of MBBA, which is known to form a liquid crystal phase below the liquid crystalline to isotropic phase transition temperature (Tc). **Figure 4B** shows the ^{1}H NMR spectrum of MBBA in the liquid crystalline state at 35°C, which is 5.5°C below the phase transition temperature (Tc = 40.5°C). A broad ^{1}H NMR spectrum with a 20 kHz linewidth was obtained for the liquid crystalline sample due to residual ^{1}H-^{1}H dipolar couplings. MBBA molecules tend to align along the magnetic field in the liquid crystalline phase; therefore, residual ^{1}H-^{1}H dipolar interactions induce a number of transitions with various degrees of dipolar interactions and this generates a significant line broadening. These dipolar interactions can provide insight into the order parameter of liquid crystals. **Figure 4C** shows a high-resolution ^{1}H NMR spectrum of MBBA in the isotropic phase that was obtained at 45°C, in which the narrow proton signals are well resolved, which enabled the assignment of the signals to their respective protons in the molecules [33].

The MBBA temperature was significantly increased by 5.0°C steps from 20.0°C below Tc to 40.5°C. As shown in **Figure 4D**, broad ^{1}H NMR signals of the liquid crystalline phase appeared alone at 35°C. At 40°C, the liquid crystalline phase had partly transitioned to the isotropic phase (**Figure 4D**). It was also evident that the liquid signals obtained at this temperature were broader than those of the fully isotropic phase, which may be attributed to the interaction of the isotropic and liquid crystalline phases, which induces a temperature distribution. This phase transition was completed at 40.5°C (**Figure 4D**), which indicates that the liquid crystalline and isotropic phases coexist near the phase transition temperature [33].

The temperature was then set at 20°C (20.5°C blow the Tc), followed by CW microwave irradiation at 130 W for 90 s, which generated weak isotropic phase

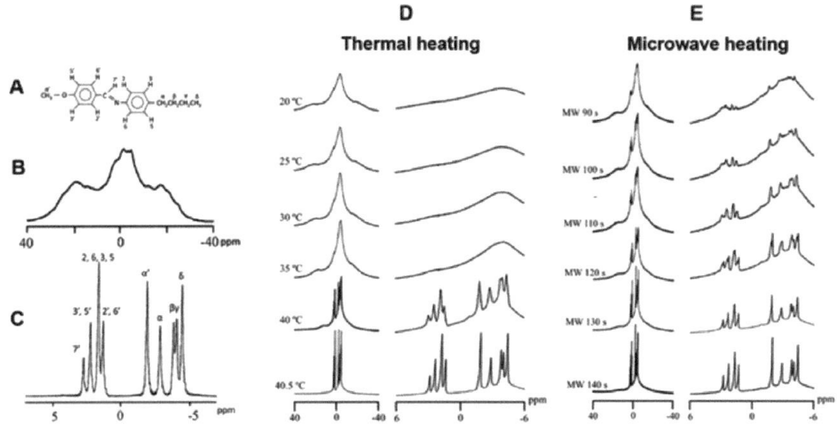

Figure 4.
A. Molecular structure of MBBA. B. ^{1}H NMR spectrum of MBBA in the liquid crystalline phase at 35°C. C. ^{1}H NMR spectrum of MBBA in the isotropic phase at 45°C, together with signal assignments for the individual protons. D. Series of ^{1}H NMR spectra of MBBA during the thermal heating processes. E. Series of ^{1}H NMR spectra during the microwave heating processes [33].

signals among the liquid crystalline phase signals (**Figure 4E**) even at a temperature lower than Tc. The temperature of the liquid crystalline phase was estimated to be 35°C from the assessment of the NMR linewidths with respect to the temperature. Such signals would not typically be expected until the temperature of the sample is close to its isotropic phase transition temperature of 40.5°C. This result indicates that microwave irradiation generated localized heating in the sample to form a small region of the higher temperature isotropic phase.

The temperature of the sample was successfully determined using *in situ* microwave irradiation NMR because the temperature of the MBBA liquid crystal was estimated with the NMR signal linewidth. It is noted that microwave irradiation generated a small fraction of the isotropic phase in the bulk liquid crystal at 35°C, even though this is 5.5°C lower than Tc (40.5°C). This result suggests non-equilibrium localized heating within the liquid crystalline sample [24].

8. Mechanism for microwave heating processes of liquid crystalline MBBA

The microwave-induced local heating phenomena observed in liquid crystalline MBBA is shown schematically in **Figure 5A**. Microwave irradiation generates a small amount of the isotropic phase inside the liquid crystalline sample below the phase transition temperature (Tc) (**Figure 5A(b)**). The dielectric loss for the isotropic phase is expected to be higher than that for the liquid crystalline phase, as shown in the MD simulation discussed in a later section. As a result, the isotropic phase is heated more efficiently by microwave irradiation, which induces a relatively higher temperature in the isotropic phase region. This phenomenon can be considered to be due to a type of non-equilibrium localized heating state, as observed in liquid–solid system [24]. The isotropic phase forms small particles and the surface of these particles interact with the surrounding liquid crystalline molecules to generate different linewidths than those produced by the bulk isotropic phase. This non-equilibrium heating state can be maintained over long time spans because the rate at which heat is obtained by the small isotropic phase particles by the absorption of microwave energy is balanced with the rate at which heat is dissipated to the bulk liquid crystalline phase. At a higher microwave power level,

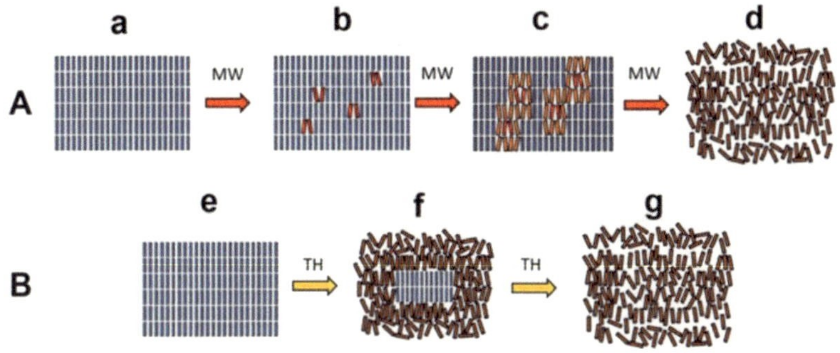

Figure 5.
A: Proposed (a, b, c and d) microwave (MW) heating processes within the liquid crystalline state. A small fraction of the liquid crystalline domain (hot spots) changes to the isotropic phase during microwave irradiation. The rate of temperature increase in this isotropic phase domain is higher than that in the liquid crystalline phase because of the larger dielectric loss for the isotropic phase. B: Schematic diagram showing the thermal heating process (e, f, and g) starting from the liquid crystalline phase to the isotropic phase [33].

the bulk isotropic phase increases (**Figure 5A(c)**), and eventually the entire sample transitions to the isotropic phase (**Figure 5A(d)**) [33].

In conventional thermal heating, the surface of the liquid crystalline state begins to melt to the isotropic state as shown in **Figure 5B(f)** and subsequently undergoes a rapid change to the isotropic state over the entire region of the sample (**Figure 5B(g)**).

9. CSC-temperature of MBBA in the isotropic phase

Figure 6A shows the structure of MBBA and the chemical shift values for individual protons as a function of temperature for isotropic phase MBBA. It is noted

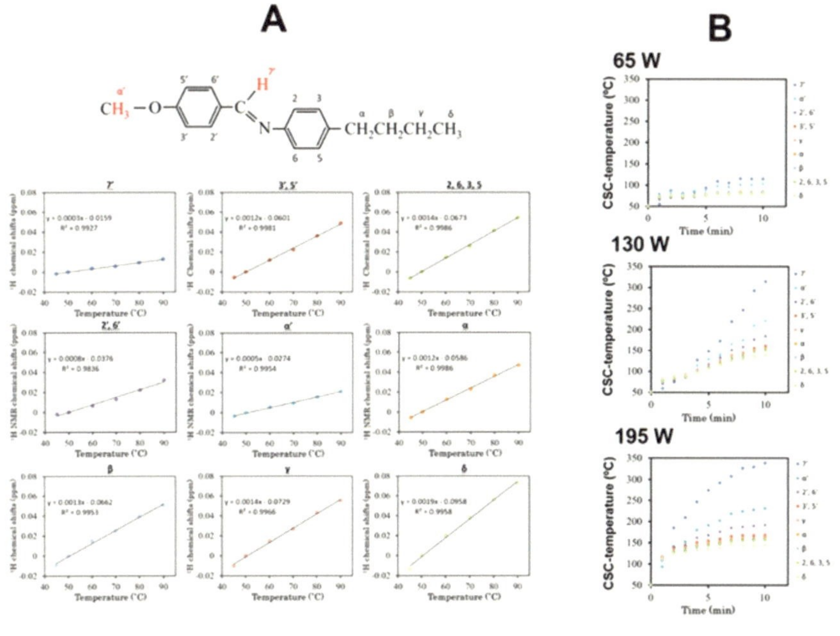

Figure 6.
A. Plots of ^{1}H chemical shift against setting the temperature of NMR spectrometer. B. Temperature as a function of the microwave irradiation time at a microwave power of 65, 130, and 195 W [33].

that the chemical shifts of different protons showed very different temperature variations. However, the chemical shift did exhibit a linear change as a function of the temperature for each different proton; therefore, it was possible to estimate the CSC-temperature of MBBA in the isotropic phase under microwave irradiation.

10. Microwave heating process of MBBA in the isotropic phase

Figure 6B presents an increase of the CSC-temperature for the MBBA sample under continuous wave (CW) microwave irradiation. The CSC-temperature increased from 50 to 70°C within 2 min under the application of 65 w CW microwave irradiation, based on the majority of protons. After microwave irradiation for 5 min, the CSC-temperature increase plateaued. However, there were significant variations in the CSC-temperatures among the protons; the 7′ and α' protons indicated 110 and 80°C, respectively. When microwave irradiation was applied at 130 W, the CSC-temperature was increased to 140°C for the majority of the protons, although values of 210 and 330°C were reached for the α' and 7′ protons, respectively. A further 8 min irradiation was required to obtain a saturated temperature. It is noted that the CSC-temperature of 7′ and α' protons showed higher CSC-temperature than the other protons. After microwave irradiation at 195 W, the CSC-temperature increased to 160°C within 5 min, although the α' and 7′ protons were discrepant from the other protons, with 220 and 350°C, respectively. Thus, α' and 7′ protons show much higher CSC-temperature than the other protons due to a non-thermal microwave effect as in the case of the OH protons of ethanol.

11. Characterization of non-thermal microwave effect of ethanol-hexane mixed solution and MBBA isotropic state

As shown in **Figure 3**, the CSC-temperature of the OH group of ethanol under microwave irradiation was considerably lower than those of the CH_2 and CH_3 protons. OH protons have a different CSC-temperature from the other protons. A temperature increase of the OH protons induces a higher field shift due to a reduction of hydrogen bonding under thermal heating [40]. Similar higher field shifts have been observed in H_2O protons [56]. In contrast, the CSC-temperatures of the OH protons are lower than the bulk temperature under microwave irradiation; therefore, the lower field shift of OH protons indicates the presence of a non-thermal microwave effect. On the other hand, the CSC-temperature of 7′ and α' protons of MBBA in the isotropic state showed much higher CSC-temperature than the other non-polar proton groups.

It is important to consider microwave heating from a physicochemical (thermodynamical) point of view to explain the energy flow mechanism. When energy is supplied to the solution under constant pressure (P) and temperature (T), the change of the Gibbs free energy (dG) is described as the difference of the changes in the enthalpy term $dH = dU + d(PV)$, where dU is the change in the total internal energy which consist of dQ(heat) and dW(work), and entropy (TdS) terms. In the case of a solution state, the change of volume (V) is very small and the d(PV) term can be neglected, so that the Gibbs free energy is given by $dG = dH - TdS = dQ + dW - TdS$. In the case of a conventional thermal heating process, the change of the Gibbs free energy (dG) mainly increases the heat energy (dQ) term, which causes a temperature increase.

Figure 7 shows the energy flow processes from a physicochemical point of view. In the thermal heating process, the directions of polar molecules fluctuate

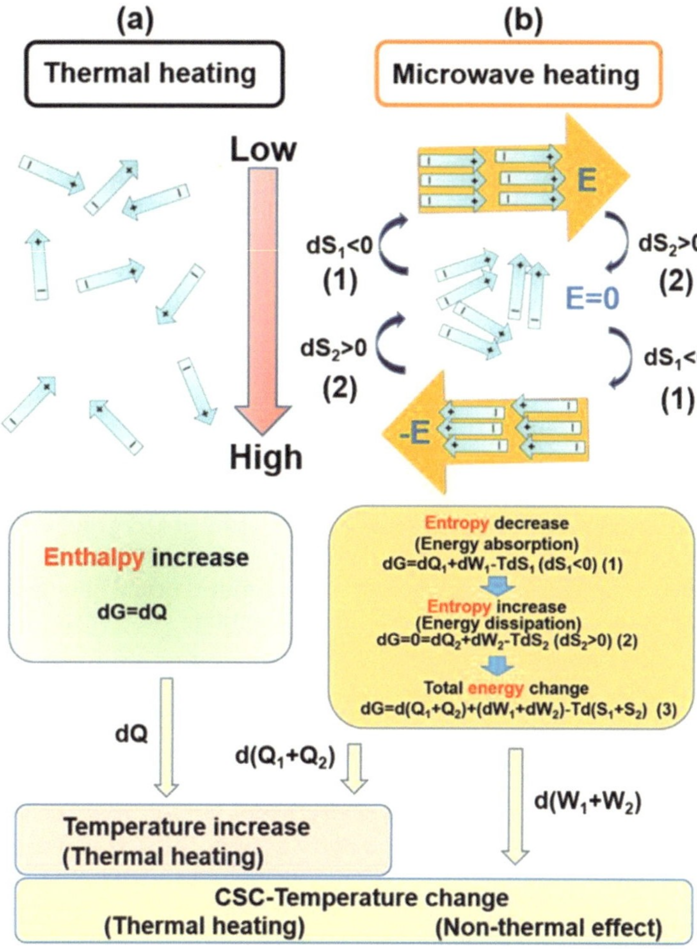

Figure 7.
Energy flow pathway under (a) conventional thermal heating and (b) microwave heating. Blue arrows indicate electric dipolar moment vectors and orange arrows indicate electric field vectors. Adapted with permission from [34]. Copyright (2020) American Chemical Society.

randomly, so that no net dipolar moment is induced (**Figure 7(a)**). In contrast, the microwave heating process induces order of the polar molecules to the electric field and a decrease of the entropy term ($dS_1 < 0$), therefore, the Gibbs free energy is increased to absorb microwave energy in the system as a first step, $dG = dQ_1 + dW_1 - TdS_1$, as shown in **Figure 7(b)**(1). The electric field oscillates with a frequency of 2.45 GHz; therefore, the polar molecular order is simultaneously reduced by an increase of the entropy term ($dS_2 > 0$) by conserving the Gibbs free energy, $0 = dQ_2 + dW_2 - TdS_2$, as a second step (**Figure 7(b)**(2)). Energy (TdS_2) thus dissipates as heat and work terms, $dQ_2 + dW_2$ to the system. By considering the sum of these two steps, $dG = d(Q_1 + Q_2) + d(W_1 + W_2) - Td(S_1 + S_2)$, (**Figure 7(b)**(3)), the heat term $d(Q_1 + Q_2)$ increases the temperature as thermal microwave effect and the work term $d(W_1 + W_2)$ may change the CSC-temperature for the OH group as a non-thermal microwave effect involving work terms such as molecular ordering and hydrogen bond formation. We interpret how the dW term changes the CSC-temperature of the OH groups in ethanol and H-C=N (7′) and CH_3-O (α’) groups in MBBA.

It is important to note that polar molecules follow the oscillating electric field in a coherent manner. In this case, coherently aligned polar molecules are able to

interact with each other; therefore, there is an electrostatic interaction between the molecules that may specifically change the electric polarization in the polar group (OH, H-C=N, and CH_3-O groups) and thus cause a change of the electron density in these group, thereby inducing a chemical shift change. OH groups are polarized to O^-H^+ in the presence of an electric field, so that the electron density of OH protons may be reduced and the ^{1}H chemical shift is therefore expected to shift to the lower field under microwave irradiation. Since this process does not change the thermal heat energy of the system, the non-thermal microwave effect of dW is evident. In the case of ethanol, molecular order may increase the number of hydrogen bonds between the OH groups because ethanol molecules form clusters in a non-polar solvent [57], which induces a lower field chemical shift due to a microwave non-thermal effect that is in the direction opposite to conventional thermal heating.
In summary, the entropy term is decreased to supply the microwave energy to the system and the entropy term is then subsequently increased because of dielectric loss by the change of electric field to dissipate the (dQ + dW) energy to the system (**Figure 7(b)**). As a result, the temperature is increased by the dQ term due to the thermal microwave effect, and the CSC-temperature of OH groups is further changed by the dW term due to the non-thermal microwave effect. MD simulation was further performed to characterize thermal and non-thermal microwave effects from a microscopic point of view in the following section.

12. Microscopic behavior of ethanol and MBBA molecules under microwave irradiation revealed by MD simulation

The microscopic behavior of ethanol in an ethanol-hexane mixed solution and MBBA molecules under microwave irradiation was further investigated using MD simulation. The results of ethanol-hexane and MBBA systems are shown in **Figure 8A(a–e)** and **B(f–h)**, respectively. Eight main simulations of the ethanol-hexane system were performed at 303, 313, 323, and 333 K in the presence and absence of an oscillating electric field. The net dipole moment induced by the oscillating electric field of 2.45 GHz is shown in **Figure 8(a)**. **Figure 8(b)** shows an ensemble of electric dipole moments along the x-direction (p_x) as a function of time at 303 K; p_x oscillated as a function of time. The phase of p_x was delayed from the electric field oscillation by a delay of around 36 ps, which was close to the experimentally measured average dielectric relaxation time of the ethanol-hexane mixture averaged around 30 ps [58]. This result is related to the dielectric constant and dielectric loss of the ethanol-hexane mixed solution, which induces the thermal microwave effect. Similar behavior was also observed in previous simulations of ice, water, and saline solutions [5]. In contrast, p_x fluctuated around zero in the absence of an electric field because of the random thermal fluctuation of the molecules in the solution.

The polar OH groups of ethanol molecules could affect the dipole moment of the system. Therefore, the orientation behavior of the OH groups was evaluated. The net direction of OH groups along the x-axis was calculated during all simulations, which is the sum of the length of all OH groups $\left(OH_x = \sum_i \left(H_{x,i} - O_{x,i}\right)\right)$. **Figure 8(c)** shows the net x-direction of OH groups during the simulations at 303 K in the presence (black) and absence (red) of the electric field E. The net direction oscillated as a function of time in the presence of an electric field, unlike the random fluctuation without an electric field. This indicates that microwave irradiation suppresses the random movement of the polar groups and causes their coherent alignment [59]. This alignment decreases the entropy of the system compare with

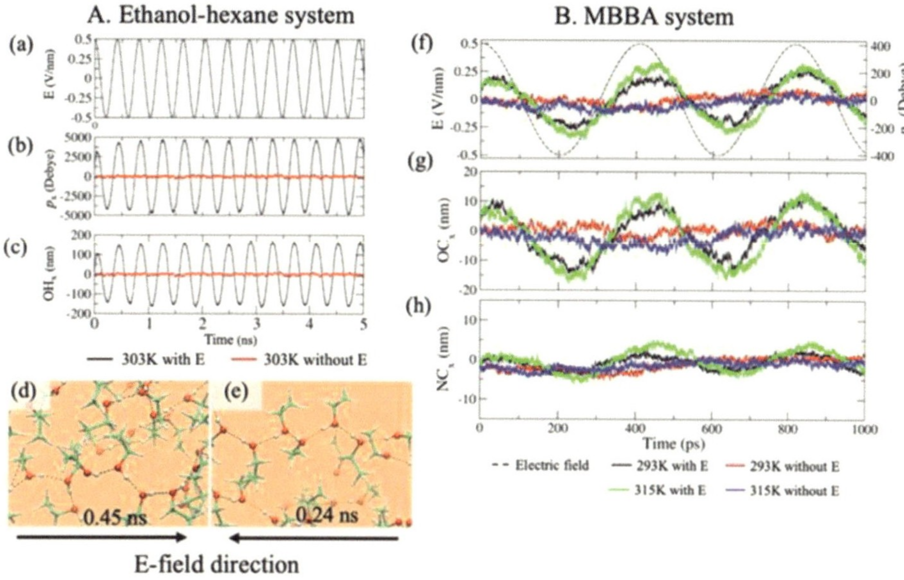

Figure 8.
A. Ethanol-hexane (1:1) system. (a) Applied electric field as a function of time. (b) the dipole moment of the simulations in the presence (black) and absence (red) of an electric field along the x-axis as a function of time at 303 K. (c) Net direction of OH groups as a function of time. (e) and (d) show examples of hydrogen bond formation in snapshots of the simulation at 0.24 and 0.45 ns under negative and positive electric field directions at 303 K, respectively. The hexane molecules were omitted for clarity. The dashed lines represent the hydrogen bonds between ethanol molecules, and the black arrows indicate the electric field directions in (d) and (e). Adapted with permission from [34]. Copyright (2020) American Chemical Society. B. MBBA system. (f) Applied electric field (dashed black) and the dipole moments of MBBA along the x-axis as a function of time. Net direction of C-O (g) and C=N (h) bonds of MBBA as a function of time. Black and green lines indicate the oscillation of these values as a function of time under the applied electric field at 293 and 315 K, respectively. Red and blue lines indicate the random fluctuations of the values without the electric field at 293 and 315 K, respectively.

that of the random orientations. **Figure 8(d)** and **(e)** show microscopic pictures of the ethanol molecules at 303 K in the presence of positive (+E) and negative (−E) electric fields at 0.45 and 0.24 ns, respectively. It is important to note that the orientation behavior of OH bonds along the electric field direction can be observed in these figures. In contrast, the directions of the OH groups were mostly oriented in random directions in the absence of an electric field [34]. Similar results were observed in other simulations at 313, 323, and 333 K (data are not shown). A time delay between the oscillation phases of the electric field and of the net OH direction was observed, which may have perturbed the ordered state of the polar groups and caused an increase in entropy due to the dielectric loss. The energy can consequently be dissipated into the system to increase the temperature via a thermal microwave effect. Furthermore, the OH groups formed a higher number of hydrogen bonds under the electric field compared to the groups without an electric field at higher temperatures [34]. This result explains the experimental data showing that the ^{1}H resonance of the OH group has a lower field shift and thus a lower CSC-temperature than the bulk temperature, as well as the lower field chemical shift increases at a higher temperature. It should be noted that this type of chemical shift induced by the electron polarization in the presence of an electric field under microwave irradiation may not appear under conventional thermal heating.

MD simulations of MBBA at 293 K (<Tc) and 315 (>Tc) K with and without the external electric field were also performed. **Figure 8(f)** shows the applied electric field (dashed black line) and the dipole moments of MBBA in these simulations along the x-axis as a function of time. The black and green lines show the oscillation of the dipole moments along the x-axis under the applied electric field at 293 and 315 K, respectively. Due to the applied electric field, the dipole moment (black and green) at 293 and 315 K were delayed by around 32.5 and 18 ps, respectively. The values are close to the experimentally measured dielectric relaxation time of MBBA at 28 and 23.7 ps at 20 and 50°C, respectively [60]. In contrast, there is no oscillation of the dipole moment without the electric field at 293 K or 315 K, which indicates that the applied electric field affected the polar groups of the MBBA molecules.

The net direction of C-O and C=N groups of MBBA during these simulations were also analyzed, and are shown in **Figure 8(g)** and **(h)**, respectively. The α' and 7' protons respectively bind to the C-O and C=N groups in the MBBA molecule, as shown in **Figure 6A**. **Figure 8(g)** and **(h)** show the net directions of the C-O and C=N groups as a function of time. The black and green lines show the net direction of C-O or C=N groups under the electric field at 293 and 315 K, respectively. The red and blue lines display the net directions in the simulations at 293 and 315 K without an electric field. The oscillation phase of the C-O groups is consistent with the phase of the dipole moments, as shown in **Figure 8(f)**, and their amplitudes are higher than those of the C=N groups shown in **Figure 8(h)**. This indicates that microwave heating has a more significant influence on the C-O groups of MBBA molecules.

These simulations showed the ordering of the polar molecules in the presence of an electric field as a non-thermal microwave effect. This polarization can only appear under microwave irradiation. Furthermore, this type of ordered state is also considered a low entropy state and is a distinct state occurring under microwave irradiation.

13. Microwave effects on chemical reaction processes

According to the MD simulations under microwave irradiation, the interactions between coherently ordered polar groups increased the number of hydrogen bonds in the ethanol-hexane mixed solution due to their coherent polarization (**Figures** 7 and **8**). The formation of hydrogen bonds is due to the interaction between two OH groups, thus causing the energy of the work term to be supplied as a non-thermal microwave effect. This non-thermal microwave effect was experimentally verified by the observation of a lower field ^{1}H chemical shift of the OH protons in ethanol. This coherently ordered state of OH groups only appears under microwave irradiation and is different from the molecular order achieved by conventional thermal heating, even at the same bulk temperature of the system. In this microwave-induced ordered state, polar molecules are coherently aligned along with the alternately oscillated electric field. These coherently ordered molecules enable interaction between polar groups. Furthermore, the coherently ordered low entropy state may accelerate the chemical reaction rate between molecules with polar groups, as in the formation of hydrogen bonding in ethanol.

Similar to under microwave irradiation, polarizable molecules, or those with a dipole moment, will also gradually align with the direction of an oriented external electric field (OEEF). A sufficiently strong OEEF can completely orient a molecule or a molecular complex in space through interacting with its dipole and polarizability, thereby removing, in principle, the difficulty in orienting the molecules and the OEEF. Therefore, it is possible to enhance or control the chemical reactivity in catalysis by a decrease in the activation energy of the reaction [61, 62].

In the case of microwave irradiation, the electric field is oscillating at a frequency of 2.45 GHz. As discussed in the MD simulation and thermodynamics consideration, a polar molecule is coherently ordered along with the oscillating electric field. Unlike in the case of an applied OEEF, the coherent ordered state under microwave irradiation alternates at a frequency of 2.45 GHz. Nevertheless, the lifetime of a coherently ordered state is sufficiently long to accelerate the chemical reaction rate.

14. Conclusion

The CSC-temperatures of an ethanol-hexane mixed solution and MBBA in the isotropic state under microwave irradiation were accurately evaluated using the linear relationship of temperature with respect to the ^{1}H chemical shift changes ($\Delta\delta$) of individual protons. A CSC-temperature increase was observed as a function of the microwave irradiation time for CH_2 and CH_3 non-polar protons. The CSC-temperature for non-polar protons reflects the bulk temperature of the solution. A lowered CSC-temperature with lower field ^{1}H chemical shift was observed for OH polar protons than that with CH_2 and CH_3 non-polar protons in ethanol, and higher CSC-temperature was observed for H-C=N (7′) and CH_3-O (α’) protons in MBBA. The lowered CSC-temperature of OH protons in ethanol under microwave irradiation which was lower than the bulk temperature is concluded to be the experimental evidence of a non-thermal microwave effect. In the microwave heating process, microwave energy is absorbed into the polar molecular system by the formation of an ordered state with lower entropy. Ordered dipolar molecules cannot completely follow the oscillating electric field; therefore, the ordered state becomes partly disordered, increasing the entropy. Microwave energy is simultaneously dissipated to the system as thermal and non-thermal microwave effects. These coherently ordered molecules interact strongly with each other to form hydrogen bonds between the OH groups of ethanol, and these interactions are considered to be due to a non-thermal microwave effect. MD simulation was carried out to confirm the theoretical validity of the experimentally observed increased lower field ^{1}H chemical shift, and the results were found to agree well. These non-thermal microwave effects play an important role in the intrinsic acceleration of chemical reaction rates between polar molecules under microwave irradiation. It is considered that the coherently ordered state reduces the activation energy for the reaction, which increases the reaction rate as catalysis.

Acknowledgements

This work was supported by KAKENHI Grant-in Aid (JP16H00756 to AN and JP20H05211 to IK) from the Ministry of Education, Culture, Sports, Science and Technology (MEXT), Japan and by KAKENHI Grant-in-Aid (JP15K06963 to AN and JP18H02387 to IK) from the Japan Society for the Promotion of Science (JSPS). The calculations were performed using clusters or supercomputers at the Research Center for Computational Science, Okazaki, Japan. The authors thank Ms. N. Yamaguchi for her financial support.

Author details

Akira Naito[1]*, Yugo Tasei[1], Batsaikhan Mijiddorj[1,2], Izuru Kawamura[1] and Kazuyoshi Ueda[1]

1 Graduate School of Engineering, Yokohama National University, Yokohama, Japan

2 School of Engineering and Applied Sciences, National University of Mongolia, Ulanbaatar, Mongolia

*Address all correspondence to: naito@ynu.ac.jp

References

[1] Horikoshi S, Schiffmann RF, Fukushima J, Serpone N. Microwave chemical and materials processing. A Tutorial. Springer Nature. **2018**; 393 p. DOI:10.1007/978-981-10-6466-1

[2] Perreux L, Loupy A. A tentative rationalization of microwave effects in organic synthesis according to the reaction medium, and mechanistic considerations. Tetrahedron. **2001**;57:9199-9223. DOI: 10.1016/S0040-4020(01)00905-X

[3] Lidström P, Tierney J, Wathey B, Westman, J. Microwave assisted organic synthesis-a review. Tetrahedron. **2001**;57:9225-9283. DOI: 10.1016/S0040-4020(01)00906-1

[4] Kappe CO. Controlled microwave heating in modern organic synthesis. Angew. Chem. Int. Ed. **2004**;43:6250-6284. DOI: 10.1002/anie.200400655

[5] Tanaka M, Sato M. Microwave heating of water, ice, and saline solution: molecular dynamics study. J. Chem. Phys. **2007**;126:034509. DOI:10.1063/1.2403870

[6] Kanno M, Nakamura K, Kanai E, Hoki K, Kono H, Tanaka M. Theoretical verification of nonthermal microwave effects on intramolecular reactions. J. Phys. Chem. A. **2012**;116:2177-2183. DOI: 10.1021/jp212460v

[7] Kappe CO, Pieber B, Dallinger, D. Microwave effect in organic synthesis: myth or reality? Angew. Chem. Int. Ed. **2013**;52:1088-1094. DOI:10.1002/anie.201204103

[8] Gedye R, Smith F, Westaway K, Ali H, Baldisera L, Laberge L, Rousell, J. The use of microwave ovens for rapid organic synthesis. Tetrahedron Lett. **1986**;*27*:279-282. DOI: 10.1016/S0040-4039(00)83996-8

[9] Giguere RJ, Bray TL, Duncan SM, Majetich G. Application of commercial microwave ovens to organic synthesis. Tetrahedron Lett. **1986**;29:4945-4948. DOI: 10.1016/S0040-4039(00)85103-5

[10] Adam D. Out of the kitchen. Nature. **2003**;421:571-572. DOI:10.1016/S0040-4039(00)85103-5

[11] Bogdal D, Lukasiewicz M.; Pielichowski J, Miciak A, Bednarz Sz. Microwave-assisted oxidation of alcohols using magtrieve. *Tetrahedron*. **2003**;*59*:649-653. DOI: 10.1016/S0040-4020(02)01533-8

[12] Yoshimura Y, Shimizu H, Hinou H, Nishimura S. A novel glycosylation concept; microwave-assisted acetal-exchange type glycosylations from methyl glycosides as donors. Tetrahedron Lett. **2005**;46:4701-4705. DOI:10.1016/j.tetlet.2005.05.046

[13] Parker M, Besson T, Lamare S, Legoy, M. Microwave radiation can increase the rate of enzyme-catalysed reactions in organic media. Tetrahedron Lett. **1996**;37:8383-8386. DOI: 10.1016/0040-4039(96)01544-4

[14] Shimizu H, Yoshimura Y, Hinou H, Hishimura, S. A new glycosylation method part 3: study of microwave effects at low temperatures to control reaction pathways and reduce byproducts. Tetrahedron. **2008**;64:10091-10096. DOI:10.1016/j.tet.2008.08.011

[15] Hoogenboom R, Wiesbrock F, Huang H, Leenen MAM, Thijs HML, van Nispen SFGM, van der Loop M, Fustin C, Jonas AM, Goby J, Schubert US. Microwave-assisted cationic ring-opening polymerization of 2-oxazolines: a powerful method for the synthesis of amphiphilic triblock copolymers. Macromelecules. **2006**;39:4719-4725. DOI: 10.1021/maq0609252a

[16] Iwamura T, Ashizawa K, Sakaguchi M. Efficient and echo-friendly anionic polymerization of acrylamide under microwave irradiation and hydrolysis of the obtained polymers by microwave irradiation. Macromolecules. **2009**;42:5001-5006. DOI: 10.1021/ma900769e

[17] Kajiwara Y, Nagai A, Chujo Y. Microwave-assisted synthesis of poly(2-hydroxyethyl methacrylate) (HEMA)/Silica hybrid using *in situ* polymerization method. Polymer J. **2009**;41:1080-1084. DOI: 10.1295/polymj.PJ2009157

[18] Yamada S, Takasu A, Takayama S, Kawamura K. Microwave-assisted solution polycondensation of L-lactic acid using a Dean-Stark apparatus for a non-thermal microwave polymerization effect induced by the electric field. Polym. Chem. **2014**;5:5283-5288. DOI: 10.1039/c4py00639a

[19] Pramanik BN, Mirza UA, Ing YH, Liu Y, Bartner PL, Weber PC, Bose AK. Microwave-enhanced enzyme reaction for protein mapping by mass spectroscopy: a new approach to protein digestion in minutes. Protein Sci. **2002**; *11*:2676-2687. DOI: 10.1110/ps.0213702

[20] Huang W, Xia, Y, Gao H, Fang Y, Wang Y, Fang Y. Enzymatic esterification between n-alcohol homologs and n-caprylic acid in non-aqueous medium under microwave irradiation. J. Mol. Cata. B Enzym. **2005**;*35*:113-116. DOI:10.1016j.molcatb.2005.06.004

[21] Lin, S, Wu C, Sun M, Sun, C, Ho Y. Microwave-assisted enzyme-catalyzed reaction in various solvent systems. J. Am. Soc. Mass Spectrom. **2005**;16:581-588. DOI: 10.1016/j.jasms.2005.01.012

[22] Herrero MA, Kremsner JM, Kappe CO. Nonthermal microwave effects revisited: on the importance of internal temperature monitoring and agitation in microwave chemistry. J. Org. Chem. **2008**;73:36-47. DOI: 10.1021/jo7022697

[23] Obermayer D, Gutmann B, Kappe CO. Microwave chemistry in silicon carbide reaction vials: separating thermal from nonthermal effects. Angew. Chem. Int. Ed. **2009**;48:8321-8324. DOI: 10.1002/anie.200904185

[24] Tsukahara Y, Higashi A, Yamauchi T, Nakamura T, Yasuda M, Baba A, Wada Y. In situ observation of nonequilibrium local heating as an origin of special effect of microwave on chemistry. J. Phys. Chem. C. **2010**;114:8965-8970. DOI:10.1021/jp100509h

[25] Naito A, Imanari M, Akasaka K. Separation of local magnetic fields of individual protons in nematic phase by state-correlated 2D NMR spectroscopy. J. Magn. Reson. **1991**;92:85-93. DOI: 10.1016/0022-2364(91)90249-S

[26] Naito A, Ramamoorthy A. Atomic-resolution Structural Studies of Liquid Crystalline Materials Using Solid State NMR Techniques. In: Ramamoorthy A , editor. Thermotropic Liquid Crystal: Recent Advances. Springer; **2007**;p.85-116. DOI: 10.1021/ja061153a

[27] Naito A, Imanari M, Akasaka K. State-correlated two-dimensional NMR spectroscopy: separation of local dipolar fields of protons in nematic phase of 4′-methoxybenzylidene-4-acetoxyaniline. J. Chem. Phys. **1996**;105:4502-4510. DOI: 10.1063/1.472300

[28] Akasaka K, Kimura M, Naito A, Kawahara H, Imanari M. Local order, conformation, and interaction in nematic 4-(n-pentyloxy)-4′-cyanobipheneyl and its one-to-one mixture with 1-(4′-cyanophenyl)-4-propylcyclohexane. A study by state-correlated 1H two-dimensional NMR spectroscopy. J. Phys. Chem.

1995;99:9523-9529. DOI: 10.1021/j100023a034

[29] Naito A, Tasei Y. Separation of local fields of individual protons in nematic phase of 4′-ethoxybenzylidene-4-n-butylaniline by microwave heating 2D NMR spectroscopy. *Mater. Sci. Technol. (M S&T)*. **2010**:2886-2894.

[30] Naito A, Makino Y, Tasei Y, Kawamura I. Photoirradiation and microwave irradiation NMR spectroscopy. In: The NMR Society of Japan, editor. Experimental approaches of NMR spectroscopy. Methodology and application of life science and materials science. Springer. **2018**;Ch 5: p. 135-170. DOI:10.1007/978-981-10-5966-7_5

[31] Akasaka K, Naito A, Imanari M. Novel method for NMR spectral correlation between the native and the denatured states of a protein. Application to ribonuclease A. J. Am. Chem. Soc. **1991**;113:4688-4689. DOI:10.1021/ja00012a052

[32] Tasei Y, Yamakami T, Kawamura I, Fujito T, Ushida K, Sato M, Naito A. Mechanism for microwave heating of 1-(4′-cyanophenyl)-4-propylcyclohexane characterized by *in situ* microwave irradiation NMR spectroscopy. J. Magn. Reson. **2015**;*254*:27-34.DOI: 10.1016/j.jmr.2015.02.002

[33] Tasei Y, Tanigawa F, Kawamura I, Fujito T, Sato M, Naito A. The microwave heating mechanism of N-(4-methoxybenzylliden)-4-butylaniline in liquid crystalline and isotropic phases as determined using *in situ* microwave irradiation NMR spectroscopy. Phys. Chem. Chem. Phys. **2015**;17:9082-9089. DOI: 10.1039/c5cp00476d

[34] Tasei Y, Mijiddorj B, Fujito T, Kawamura I, Ueda K, Naito A. Thermal and nonthermal microwave effects of ethanol and hexane mixed solution as revealed by in situ microwave irradiation nuclear magnetic resonance spectroscopy and molecular dynamics simulation. J. Phys. Chem. B. **2020**;147:9615-9624. DOI:10.1021/acs.jpcb,0c06383

[35] Caleman C, van der Spoel, D. Picosecond melting of ice by an infrared laser pulse: A simulation study. Angew. Chem. Int. Ed. **2008**;*47*:1417-1420. DOI:10.1002/anie.200703987

[36] Marklund EG, Ekeberg T, Moog M, Benesch JLP, Caleman C. Controlling protein orientation in vacuum using electric fields. J. Phys. Chem. Lett. **2017**;8:4540-4544. DOI: 10.1021/acs.jpclett.7b02005

[37] English NJ, MacElroy JMD. Molecular dynamics simulations of microwave heating of water. J. Chem. Phys. **2003**;118:1589-1592. DOI: 10.1063/1.1538595

[38] English NJ. Molecular dynamics simulations of microwave effects on water using different long-range electrostatics methodologies. Mol. Phys. **2006**;104:243-253. DOI: 10.1080/14733140500352322

[39] Garcia-Barnos B, Reinosa JJ, Penaranda-Fox FL, Fernandez JF, Catala-Civera JM. Temperature assessment of microwave-enhanced heating processes. Sci. Reps. **2019**;*9*:10809. DOI: 10.1038/s41598-019-47296-0

[40] Van Geet A. Calibration of the methanol and glycol nuclear magnetic resonance thermometers with a static thermistor probe. Anal. Chem. **1968**;40: 2227-2229. DOI: 10.1021/ac50158a064

[41] Van Geet AL. Calibration of methanol nuclear magnetic resonance thermometer at low temperature. Anal. Chem. **1970**;42:679-680. DOI:10.1021/ac60288a022

[42] Zuo CS, Metz KR, Sun Y, Sherry AD. NMR temperature measurements using a paramagnetic lanthanide complex. J. Magn. Reson. **1998**;133:53-60. DOI:10.1006/jmre.1998.1429

[43] Bielecki A, Burum DP. Temperature dependence of ^{207}Pb MAS spectra of solid lead nitrate. An accurate sensitive thermometer for variable-temperature MAS. J. Magn. Reson, Ser. A **1995**;116:215-220. DOI: 10.1006/jmra.1995.0010

[44] Hoffman RE, Becker ED. Temperature dependence of the ^{1}H chemical shift of tetramethylsilane in chloroform, methanol, and dimetylsulfoxide. J. Magn. Reson. **2005**;176:87-98. DOI: 10.1016/j.jmr.2005.05.015

[45] Abraham MJ, van der Spoel D. Lindahl E, Hess B. The GROMACS development team, GROMACS User Manual version 2018.7, *www.gromacs.org, 2019*.

[46] Huang J, MacKerell Jr AD. CHARMM36 all-atom additive protein force field: Validation based on comparison to NMR data. J. Comput. Chem. **2013**;34: 2135-2145. DOI: 10.1002/jcc.23354

[47] Vanommeslaeghe K, Hatcher E, Acharya C Kundu S, Zhong S, Shim J, Darian E, Guvench O, Lopes P, Vorobyov I, Mackerell Jr AD. CHARMM general force field: A force field for drug-like molecules compatible with the CHARMM all-atom additive biological force fields, J. Comput. Chem. **2010**;31:671-690. DOI: 10.1002/jcc.21367

[48] Harish Vagadia B, Vanga SK, Singh A, Raghavan V. Effects of thermal and electric fields on soybean trypsin inhibitor protein: A molecular modelling study. *Innov.* Food Sci. Emerg. Technol. **2016**;35:9-20. DOI:10.1016/j.ifset.2016.03.004

[49] Bussi G, Donadio D, Parrinello M. Canonical sampling through velocity rescaling. J. Chem. Phys. **2007**;*126*:014101. DOI: 10.1063/1.2408420

[50] Parrinello M, Rahman A. Polymorphic transitions in single crystals: A new molecular dynamics method. J. Appl. Phys. **1981**;52:7182-7190. DOI:10.1063/1.328693

[51] Darden T, York D, Pedersen L. Particle mesh Ewald: An N·log(N) method for Ewald sums in large systems. J. Chem. Phys. **1993**;98:10089-10092. DOI:10.1063/1.464397

[52] Essmann U, Perera L, Berkowitz ML, Darden T, Lee H, Pedersen LG. A smooth particle mesh Ewald method. J. Chem. Phys. **1995**;103:8577-8593.

[53] Hess B, Bekker H, Berendsen HJC, Fraaije JGEM. LINCS: A linear constraint solver for molecular simulations. J. Comput. Chem. **1997**;18:1463-1472. DOI: 10.1002/(SICI)1096-987X(199709)18:12<1463::AID-JCC4>3.0.CO;2-H

[54] Vaught A. Graphing with Gnuplot and Xmgr: Two graphing packages available under Linux. Linux J. **1996**;1996:7. DOI: 10.5555/326327.326334

[55] Humphrey W, Dalke A, Schulten K. VMD: Visual molecular dynamics. *J.* Mol. Graphics. **1996**;*14*:33-38. DOI: 10.1016/0263-7855(96)00018-5

[56] Dvinskikh SV, Yamamoto K, Dürr, UHN, Ramamoorthy A. Sensitivity and resolution enhancement in solid-state NMR spectroscopy of bicelles. J. Magn. Reson. **2007**;184:228-235. DOI: 10.1016/j.jmr.2006.10.004

[57] Sumi T, Dillert, R, Horikoshi S. Novel microwave thermodynamics

model for alcohol with clustering structure in nonpolar solution. J. Phys. Chem. B. **2015;**119:14479-14485. DOI: 10.1021/acs.jpcb.5b06168

[58] Utzel H, Stockhausen M. Dielectric Relaxation in Binary and Ternary Mixtures of Ethanol, Water, and Benzene or n-Hexane. Z. Naturforsch. **1985**;40a:588-595. DOI: 10.1515/zna-1985-0609

[59] Zhang Y–M, Li J-L, Wang J-P, Yang X-S, Shao W, Xiao S-Q, Wang -Z. Research on epoxy resin decomposition under microwave heating by using ReaxFF molecular dynamics simulations. RSC Adv. **2014**;*4*:17083-17090. DOI:10.1039/C4RA00473F

[60] Arora VP, Agarwal VK. Dielectric Relaxation Studies of N-(p-Methoxybenzylidene)-p-Butylaniline and N-(p-Ethoxybenzylidene)-p-Butylaniline in Benzene at Microwave Frequencies, J. Phys. Soc. Jpn. **1977**;42:908-910. DOI: 10.1143/JPSJ.42.908

[61] Shaik S, Ramanan R, Danovich D, Mandal D. Structure and Reactivity/Selectivity Control by Oriented-External Electric Fields. Chem. Soc. Rev. **2018**;14:5125-5145. DOI: 10.1039/C8CS00354H

[62] Shaik S, Danovich D, Joy J, Wang Z, Stuyver T. Electric-Field Mediated Chemistry: Uncovering and Exploiting the Potential of (Oriented) Electric Fields to Exert Chemical Catalysis and Reaction Control. J. Am. Chem. Soc. **2020**;142:12551-12562. DOI: 10.1021/jacs.0c05128